High-value Crops and Marketing

This Study was undertaken by International Food Policy Research Institute [IFPRI][1] in close consultation with India Resident Mission, Asian Development Bank[2]

Asian Development Bank
India Resident Mission (INRM)
4, San Martin Marg, Chanakyapuri,
New Delhi 110021, India

---

1. Bart Minten, Yashodhan Ghorpade, Anneleen Vandeplas and Ashok Gulati
2. Pratima Dayal, Principal Economist and Jaya Chatterji, Project Implementation Officer, India Resident Mission, Asian Development Bank

High-value Crops and Marketing

STRATEGIC OPTIONS FOR DEVELOPMENT IN UTTARAKHAND

ADB Asian Development Bank

IFPRI International Food Policy Research Institute

ACADEMIC FOUNDATION
NEW DELHI

First published in 2009 by

Academic Foundation
4772-73 / 23 Bharat Ram Road, (23 Ansari Road), Darya Ganj, New Delhi - 110 002 (India).
Phones : 23245001 / 02 / 03 / 04. Fax : +91-11-23245005. E-mail : academic@vsnl.com
www : academicfoundation.com

Published in association with :

Asian Development Bank

Cataloging in Publication Data--DK
Courtesy: D.K. Agencies (P) Ltd. <docinfo@dkagencies.com>

High-value crops and marketing : strategic options for development in Uttarakhand.
p. cm.
Includes bibliographical references (p.).
ISBN 13: 9788171887583
ISBN 10: 8171887589

1. Cash crops--India--Uttaranchal--Marketing. 2. Horticultural crops--India--Uttaranchal--Marketing. 3. Farm produce--India--Uttaranchal--Marketing. 4. Business logistics--India--Uttaranchal.

DDC 381.41095451 22

Typeset by Italics India, New Delhi.
Printed and bound in India.

Contents

List of Tables, Figures and Boxes

TABLES

Main Text

Annex

FIGURES

Main Text

BOXES

Foreword

Uttarakhand has the ecological advantage of multiple agro-climatic zones that can produce high-value crops and the locational advantage of proximity to markets. How well the agriculture value chain is developed and how efficiently farmer producers can be linked to lucrative markets in a fair, transparent and profitable manner is vital to the health of its economy. The Asian Development Bank (ADB) commissioned International Food Policy Research Institute (IFPRI) to undertake a study to understand the status, issues and the options for high-value crops and their marketing. It is our hope that this study will contribute towards formulation of appropriate policies and support environment in the state.

The study describes the current context of the national and international agricultural markets and the opportunities and challenges for development of high-value crops in the State. It undertook primary surveys of the functioning of the wholesale and retail markets. The value chains of some high-value crops, of importance to the State, including horticultural and floricultural crops, organic crops, seeds and medicinal and aromatic herb were studied. The study presents a cogent and practical set of recommendations as a way forward.

I wish to record our sincere appreciation to the study team comprising Bart Minten, Yashodhan Ghorpade, Anneleen Vandeplas and Ashok Gulati of IFPRI, and Pratima Dayal, Principal Economist, and Jaya Chatterji, Project Implementation Officer, Agriculture and Natural Resources, of India Resident Mission, ADB. The study was conducted under ADB's Technical Assistance (TA) for Knowledge Management and Capacity Building (TA 4780) funded by the Department for International Development (DFID) Trust Fund of the United Kingdom. We gratefully acknowledge the support of DFID.

I would like to thank the Chief Secretary of the Government of Uttarakhand and his officials of various Departments who liberally shared their understanding of the issues in the sector and guided the team effectively. I would also like to note our sincere appreciation to many others including Donors, the private sector, NGOs and farmers who have enriched the study by their contributions.

Tadashi Kondo

Country Director

India Resident Mission, ADB

Abbreviations and Acronyms

AEZ	Agri-Export Zones
APEDA	Agricultural and Processed Food Products Export Development Authority
APL	Above Poverty Line
APMC	Agricultural Produce Marketing Committee
BPL	Below Poverty Line
CA	Commission Agent
CC	Collection Centre
CITES	Convention on International Trade in Endangered Species
DASP	Diversified Agriculture Support Project
DRDA	District Rural Development Agency
EU	European Union
FD	Forest Department
FDI	Foreign Direct Investment
FRLHT	Foundation for Revitalisation of Local Health Traditions
FMCG	Fast Moving Consumer Goods
GDP	Gross Domestic Product
GMVN	Garhwal Mandal Vikas Nigam
GoI	Government of India
GoUA	Government of Uttarakhand
GSDP	Gross State Domestic Product
GVO	Gross Value of Output
HARC	Himalayan Action Research Centre
HP	Himachal Pradesh
HRDI	Herbal Research and Development Institute

HYV	High Yielding Variety
ICS	Internal Control System
IFAD	International Fund for Agricultural Development
IFOAM	International Federation of Organic Agriculture Movements
IFPRI	International Food Policy Research Institute
IPM	Integrated Pest Management
J&K	Jammu & Kashmir
KMVN	Kumaon Mandal Vikas Nigam
KVK	Krishi Vigyan Kendras
MAP	Medicinal and Aromatic Plants
MD	Mother Dairy
MOFPI	Ministry of Food Processing Industries
MPCA	Medicinal Plants Conservation Area
MT	Metric Tonne
MWS	Micro Water Shed
NDDB	National Dairy Development Board
NGO	Non Governmental Organisation
NHB	National Housing Bank
NSDP	Net State Domestic Product
NSSO	National Sample Survey Organisation
RBH	Rural Business Hub
SKU	Stock Keeping Unit
TDC	Tourism Development Company Limited
UAFDC	Uttarakhand Forest Development Committee
UDWDP	Uttarakhand Decentralised Watershed Development Programme
ULDB	Uttarakhand Livestock Development Board
UNCTAD	United Nations Conference on Trade and Development
UOCA	Uttarakhand Organic Certification Agency
UOCB	Uttarakhand Organic Commodities Board
UP	Uttar Pradesh
VPKAS	Vivekananda Parvatiya Krishi Anusandhan Sansthan

1 Introduction

The Indian agri-food system is undergoing a major transformation. First, organised retail, giving consumers a wider choice of goods, more convenience, a better shopping environment, and often lower prices, is rapidly growing. It is expected that the benefits of this trend should percolate to the mass of Indian consumers. Second, changes in consumption patterns are driving fast changes in the production basket. The production basket is diversifying in favour of high value commodities (fruit, vegetables, livestock and fisheries), in response to these changing consumption tastes and preferences. This has set the stage for different handling, storage and distribution networks.

However, the front-end has been changing differently than the back-end of the agri-food system. While the retailers, processors, logistics suppliers and others are seemingly scaling up very fast, the back-end is consistently fragmenting (Figure 1.1). It seems there is a movement towards a "consolidating top" and "fragmenting bottom" (Gulati *et al.*, 2008). For example, the average size of landholding in India has declined from 2.2 hectares in 1970-71 to 1.06 in 2003 and the number of farmers with less than 2 hectares of land has increased to 87.6 per cent operating nearly 43.5 per cent of the operated area. Hence, the question arises as to how the inertia that is generated by the rapidly growing front end players (retail, processing, and logistics) can be utilised to provide a major pull to the small and fragmenting farm sector. It thus seems that growth with inclusiveness is the challenge for the agricultural policy agenda in this area.

Figure 1.1 Dynamics of the Modern Agri-food System

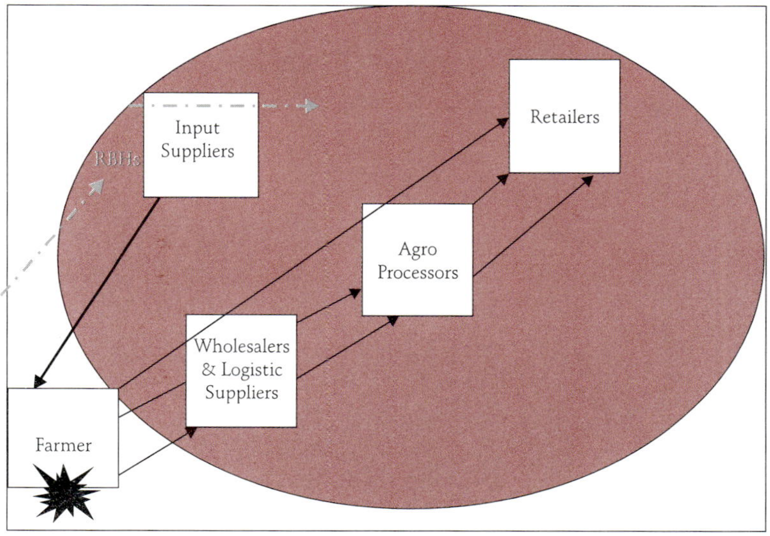

Source: Gulati *et al.* (2008).

A traditional agricultural value chain is typically crowded by many players and the layers of intermediaries that operate all along the chain result in weak farm-firm linkages. Weak linkages and poor coordination result in post-harvest losses, low value addition and thus low incomes of the farmers (Gulati *et al.*, 2008). In addition to this, a long chain of middlemen and commission agents lowers the margins accruing to each of the players in the value chain with the farmers suffering the most. In a modernising marketing system, the thrust is on developing direct farm-firm linkages and enhancing value addition along the supply chain. It seems that the competition of more efficient players with the traditional system is not only beneficial for the farmers but also for the players outside the organised value chain. However, it might not be possible to achieve these benefits in a business-as-usual environment. These changes will require policy responses and institutional reforms that create the right incentives for larger investments.

In an organised framework, the farmers can directly link with the input suppliers, wholesalers, logistics suppliers, food processors and retailers. The challenge lies in forging these backward linkages (to deliver inputs and services to the farmers). It is important to dovetail the procurement services with backward linkages through buyback guarantees, as the latter play a significant role in ensuring the sustainability of farm-firm ventures. There are various possible linkages that would include smallholders in these newly emerging value chains, i.e., farmers tying up directly with the retailers (especially for fresh produce, irrespective of the farm size), linking up with processors, specialised wholesalers, logistics suppliers, each of whom are directly linked with the modern retailers. A cluster approach is desirable to overcome the farm size problem and also ensure maximum outreach. This is possible. India has a long history of dairy cooperatives tying up with a large number of small and marginal farmers and thereby linking farmers with the markets. These trends are now also emerging in contract farming arrangements in fresh fruits and vegetables and in the poultry sector. Private retailers and processors are also starting to link with farmers directly.

Failure to take advantage of the unfolding revolution will dampen the future growth momentum of organised retail in food and grocery in India, leaving the majority of farmers in the subsistence trap and consumers no better off. It is important that the benefits of changing retail practices are not only measured by just counting the winners and losers but one has to take into account the overall impact in terms of revenue and employment generated and the long term impact on the different stakeholders in the agri-food system.

In this report, we study the case of high-value agriculture in the state of Uttarakhand in the context of all these fast marketing changes at the national and international level. The state of Uttarakhand is characterised by a significant number of opportunities in high-value agriculture. They include most importantly the presence of a high number of endemic crops, diversity in agro-climatic conditions, possibilities to produce for 'off-season' markets, organic production practices, the relative high education of producers, a strong agricultural research capacity, an active civil society, a competitive production environment and a location relatively close to terminal consumer markets, at least for part of the state (the plains and some of the mid-mountains). On the other hand, agriculture in Uttarakhand faces also significant challenges that limit the competitiveness of its farmers with farmers in other Indian states and outside India. These include the high number of small scattered farms creating problems of aggregation and transport costs, migration and land conversion, increasing water and climatic change problems, environmental vulnerability, wildlife attacks, and a problematic regulatory environment. We look at these problems in a holistic manner and suggest ways on how Uttarakhand can prepare itself better to take advantage of the changing agricultural marketing environment.

The structure of the report is as follows. In Chapter 2, we give some background information on the state and discuss the overall geography, land use, and economic performance. In Chapter 3, we present a summary view of where agricultural markets are headed at the national and international level. Chapter 4 then looks at the current state of high-value crop agriculture and the challenges and opportunities for high-value crop development. Chapter 5 documents more in particular the marketing of high-value crops, looking at the legal framework as well as effective practices. Chapter 6 presents a summary of the value chain work that is presented in more detail in the annexes I to V. We finish with recommendations in Chapter 7.

2 Understanding the Setting and Context

2.1. Introduction

With the passage of the UP State Reorganisation Bill, 2000, Uttaranchal (subsequently rechristened Uttarakhand) was established as the 27th State of the Republic of India. This represented recognition of a long-standing demand of the people of this region for a separate state to address their specific needs and concerns based on territorial, geographical and cultural distinction from the rest of Uttar Pradesh, to which it previously belonged. The creation of a new state poses both challenges and opportunities for development, and it is imperative to understand the physical and institutional environments to best gauge how opportunities may be tapped into, and challenges met.

2.2. Physical Features and Geography

Uttarakhand has strategic borders with Nepal in the East and the People's Republic of China in the North. The states of Himachal Pradesh and Uttar Pradesh lie towards the West and South respectively. The Himalayas act as a formidable natural border in the North and the generally high altitude (11 of the 13 districts are at least partly mountainous) gives the state a unique temperate climate in a tropical latitude situation. Cradled in the foothills of the Himalayas (see Figure 2.1), Uttarakhand's terrain and agro-climatic conditions present great and unique challenges as well as opportunities for growth and development.

Broadly, the state may be divided into 4 distinct agro-climatic regions:

- Valley (Up to 1000 metres above sea level): This region comprises the districts of Dehradun (parts of), Udham Singh Nagar and Haridwar.
- Low Hills (1000–1500 metres above sea level): Marked by largely un-irrigated stretches of sub-tropical land, including districts (and areas of) Pauri, Tehri, Dehradun and Nainital.
- High Hills (1500–2500 metres): This region marked by cool, temperate climate includes areas in Uttarkashi, Nainital, Rudraprayag, Pauri, Bageshwar, Champawat and Chamauli.
- Alpine Zone (Above 2500 metres): This high altitude region is marked by sparse vegetation, mainly herbs and small mountainous plants. Areas belonging to Uttarkashi, Chamoli and Pithoragarh districts comprise the alpine zone of Uttaranchal.

These zones are represented and shown in the physical map of Uttarakhand given in Figure 2.1.

The terrain of Uttarakhand, ranging from the marshy terai plains in the South to the rugged slopes in the North offers scope for varied patterns of land-use. Table 2.1 captures the land-use patterns in

Figure 2.1 **Physical Map of Uttarakhand**

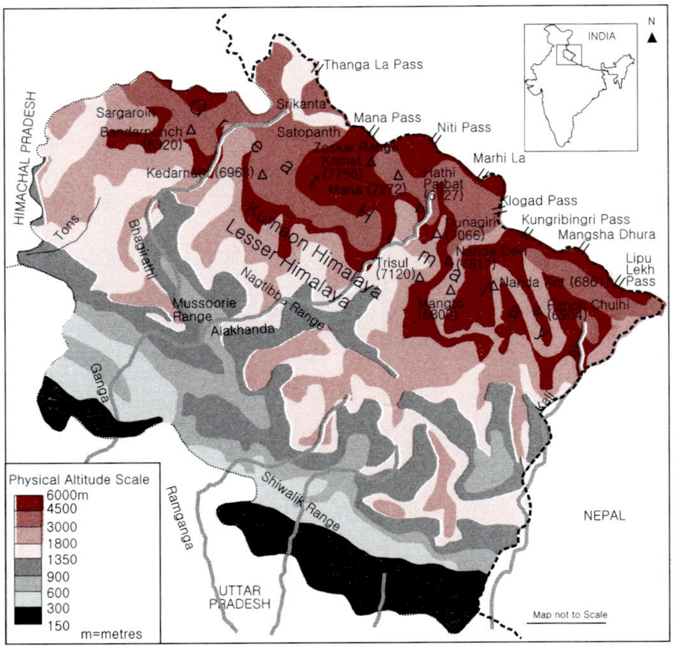

Source: www.mapsofindia.com

Table 2.1 **Land-use in Uttarakhand**

Land-use activity	Area in ha.	As % of total area
Forests	3.465.057	61
Area under non-agricultural uses	152.247	3
Barren and unculturable land	310.244	5
Culturable wasteland	384.928	7
Pastures and land under misc. tree crops and groves	481.832	8
Total fallow land	107.446	2
Net area sown	769.944	14

Source: Government of Uttarakhand (GoUA) (2007a).

the state. The table shows that the net sown area is a mere 14 per cent of the total area of the state. The area under pastures, grazing land and miscellaneous trees (not included in net sown area) accounts for 8 per cent of the total area, and is of considerable importance in higher altitudes with limited vegetation and where local communities largely depend on these for subsistence needs.

Forests are a very important part of the state's ecological setting and natural resource base, accounting for as much as 61 per cent of the total area of the state. Forests as a proportion of total area account for a smaller share over the years. The share has declined marginally from 65 per cent in 1980-81 to 64 per cent in 1993-94, and further to 61 per cent for 2001-2002 (GoUA, 2007; Joshi *et al.*, 1999).

2.3. Some Demographic, Social and Economic Indicators

Uttarakhand is a largely rural state as according to the 2001 Census, about 74 per cent of the population lives in rural areas. A look at the trends of Net State Domestic Product (NSDP) since the state's inception reveals the broad growth trajectory the state is on (Table 2.2). While the state shows remarkable growth rates (10 per cent or higher), the per capita income in the state in the year 2003-04 was Rs. 20,785, still falling marginally short of the national average of Rs. 20,989 and the poverty headcount ratio

Table 2.2 **Net State Domestic Product (NSDP) for Uttarakhand: 1999-2000 to 2004-05**

Year	Net state domestic product (Rs. Lakhs)	Annual growth rate (%)	Per capita income (Rs.)
1999-2000	1126403	-	13609
2000-01	1259811	11.8	14951
2001-02	1384342	9.9	16138
2002-03	1631150	17.8	18705
2003-04	1842822	13.0	20785
2004-05	2077027	12.7	23042

Source: GoUA (2007b).

was estimated in 2004-05 at 40 per cent, significantly higher than the national rate of 27.5 per cent.

Table 2.3 Sector-wise Composition of NSDP: Uttarakhand

in percentage

Year	Primary sector	Secondary sector	Tertiary sector
1999-2000	32.54	15.56	51.90
2000-01	31.17	16.92	51.91
2001-02	28.30	18.11	53.58
2002-03	26.77	22.11	51.13
2003-04	26.86	21.16	51.98
2004-05	24.78	23.81	51.41

Source: GoUA (2007b).

We look at the components of the NSDP to get an idea of the sectors which are generating this growth, and the ones that are lagging behind (Table 2.3). We observe that services (the tertiary sector) are the largest component of the NSDP (accounting for nearly half the state's income), distantly followed by the primary and secondary sectors. The primary sector which in 1999-2000 accounted for twice as much as the secondary sector of the share of the NSDP currently makes only a marginally higher contribution than the secondary sector. Over the years since the creation of Uttarakhand, the share of the primary sector in NDSP has fallen from 33 per cent to 25 per cent, whilst that of the secondary sector has risen from a low 16 per cent to 24 per cent. The share of services has remained more or less constant around the 51 per cent mark. One must however note that large shares of these services are not high-value (tourism being perhaps a most notable exception) and largely exist in the informal economy.

The underdevelopment of the rural areas has led many natives to migrate to bigger cities in search for employment. This migration often takes on a gendered pattern, with women typically staying back in the villages. Trends in education have been rather encouraging over the years for Uttarakhand. The literacy rate rose from 46 per cent in 1981 to 58 per cent in 1991, and was estimated to be 72 per cent in 2001, far in excess of the national average of 65 per cent, and is the second highest in North India, only after Himachal Pradesh.

The provision of basic facilities such as health, sanitation and electricity is paramount for a good quality of life, as well as indirectly through spillovers these have on workers' productivity. With regard to electricity, urban areas have a good provision, with 90 per cent of houses having connections. Rural areas however still have much catching up to do, with only about half the households having connections. Sanitation facilities in general are bad, with a mere 45 per cent households having access to these for the state as a whole. The rural areas' average is 32 per cent, while in urban areas it is 87 per cent, indicating a huge urban-rural gap.

Drinking water is a large problem in the state, and with drying up water bodies (see Krishna, 2002; Kumar, 2006) there is a great felt need for better provision of potable water in households. Currently the extent of provision is particularly worrisome in Rudraprayag and Almora districts where fewer than 20 per cent of households have access to potable water in their residential premises. This merits much concern, especially as the issue of water provision is strongly linked with environmental sustainability issues. Much of the state depends on the rains for agriculture. For instance, in Kumaon it is estimated that over 80 per cent of the crops are rainfed (Murthy and Kadekodi, 2000). Due to topographical conditions, between 20 to 35 per cent of rainfall from agricultural lands, and up to 50 per cent from community lands, accounts for surface runoff (Krishna, 2002). The absence of effective water harvesting projects, as well as declining precipitation over the years further aggravates the situation of acute water shortage. The runoff also results in loss of precious top soil, thereby reducing soil fertility and productive capacity.

3 Where are Agricultural Markets Headed?

3.1. Changes in Food Demand

Dietary habits are changing rapidly in India. For more than two decades, India has registered a growth rate of 5 to 6 per cent a year. This has increased even further in the last years. This high growth rate is strongly changing the food consumption patterns, away from basic staples and towards high-value products such as fruits and vegetables, dairy, poultry and fishery products (Joshi *et al.*, 2007). Table 3.1 shows how expenditures have changed over the last two decades based on figures from nationally representative data. First, food consumption as a part of total consumption has fallen dramatically from 66 per cent (59 per cent) in 1983 to 55 per cent (42 per cent) in rural (urban) areas. Within food expenditures, expenditures on cereals have fallen rapidly. In rural areas, cereals accounted for half of the expenditures in 1983 but fell to one third of these expenditures in 2005. Similar changes in consumption patterns are taking place in urban areas.

Table 3.1 Expenditures on Food and Food Items as Percentage of Total Consumer Expenditures in India

Food items	Expenditure as % of total consumer expenditure			
	Rural		Urban	
	1983	2004-05	1983	2004-05
Cereal	32.5	18.1	19.5	10.1
Pulses & products	3.5	3.1	3.2	2.1
Edible oil	4.0	4.6	4.8	3.5
Vegetables	4.7	6.1	5.0	4.5
Fruits & nuts	1.4	1.9	2.1	2.2
Milk & products	7.5	8.5	9.2	7.9
Meat, egg, fish	3.0	3.3	3.6	2.7
Total food	65.6	55.0	59.1	42.5

Source: GoI (2006). NSS 61st Round, Report No 508.

While we only look at percentages of total consumption, part of the explanation could be changes in (relative) prices. As to make the picture complete, we therefore look at the quantities consumed. A similar trend is emerging. The quantities consumed of cereals and of pulses are decreasing while the quantities consumed of high-value products are on the rise. Interestingly, this shift is also happening in the lower income brackets of the Indian population. For example, comparing consumption patterns of the poorest and richest segments of the population (Table 3.2), it is found that consumption of cereals is going down for both categories. While the quantities consumed of cereals declined by 20 per cent for the upper expenditure group, it also showed a significant decline of 10 per cent for the lower expenditure group. The consumption of fruits, vegetables, milk and meat, eggs and fish show dramatic increases over the last two decades and this for rich and poor alike.

Table 3.2 Dynamics of Food Consumption of Poor and Rich in India

	Lower expenditure group (30%)			Upper expenditure group (30%)		
	1983	1999-2000	% Change	1983	1999-2000	% Change
Cereal	147.1	132.4	-10.0	194.3	154.6	-20.4
Pulses	7.6	6.9	-9.2	17.7	16.6	-6.2
Edible oil	2.6	4.6	76.9	7.3	13.7	87.7
Vegetables	36.0	53.9	49.7	65.2	90.8	39.3
Fruits	1.6	4.2	162.5	6.4	18.2	148.4
Milk	15.7	20.5	30.6	89.7	117.2	30.7
Meat, egg, fish	1.9	3.8	100.0	4.8	10.6	120.8

Source: GoI (2006). NSS 61st Round, Report No 508.

Another segment where significant growth is expected because of the growing economy is within processed food. From the lower to the highest expenditures group, the share of expenditures dedicated to beverages and processed food products increases significantly and goes from 6 per cent to 24 per cent (NSSO of 1999-2000). With high income growth rates, this sector will thus boom. Food processing currently contributes around 6 per cent of the GDP in India, 6 per cent of the total industrial investment and 13 per cent of the country's exports. Although India is the world's second largest producer of many agricultural products such as fruits and vegetables, next only to the People's Republic of China, an insignificant part of it is processed. The extent of value addition is estimated at 7 per cent, compared to 45 per cent in China. A large part of the food processing in India has so far been primary in nature (e.g. rice milling) and is undertaken in the unorganised small manufacturing sector using traditional technologies.

Detailed data on current consumption habits and on the consumers can be used to draw inferences on the future demands given assumptions on the characteristics of the population. Mittal (2006) makes projections, based on estimates from demand elasticities from the current consumption data and based on assumptions on economic growth, to evaluate food consumption levels and food composition in 2020 in India (Table 3.3). She finds that total cereal consumption will increase by 2.1 per cent between 2000 and 2020. This compares to 4.5 per cent for fruits and vegetables and 6.4 per cent for milk. In quantity terms, milk is expected to become more important than cereals while fruits and vegetables will reach 85 per cent of the cereal consumption level in 2020. In value

Table 3.3 Projected Annual Domestic Demand of Food in India (Assume GDP Growth of 7%)

	Per capita (kg/capita)			Total (million tonnes)		
	Base year	Projection		Base year	Projection	
	2000	2010	2020	2000	2010	2020
Cereals	138.98	148.87	160.28	142.7	175.5	215.7
Pulses	13.27	15.99	20.21	13.6	18.8	27.2
Fruits and vegetables	74.87	97.92	135.91	76.9	115.4	182.9
Milk	64.91	100.37	170.88	66.7	118.3	229.9
Edible oil	8.89	10.89	13.98	9.1	12.8	18.8
Sugar	12.41	16.8	24.32	12.7	19.8	32.7
Meat, fish, eggs	6.49	10.46	18.74	6.7	12.3	25.2

Source: Mittal (2006).

terms, these high-value products will become significantly more important given their high value per kg.

Changes at the consumption levels have obvious implications on agricultural activities and areas planted in India. Table 3.4 shows the dynamics in agricultural output between 1982/83 and 2002/03. The share of crops in total agricultural output is still an overwhelming 71 per cent but its share is declining. It decreased from 77 per cent in 1982/83 to 71 per cent in 2002/03. Most of this decline is due to the increasing importance of livestock. Overall, the share of high-value agriculture has increased from 32 per cent in 1982/83 to 44 per cent in 2002/03.

Table 3.4 Dynamics in Indian Agriculture between 1982/83-2002/03

Commodity	Share in gross value of agricultural output (%)			Compound annual growth (%)	
	TE 1982/83	TE 1992/93	TE 2002/03	1980/81 to 1991/92	1992/93 to 2002/03
Crops	77.3	74.3	70.9	2.8	2.5
Foodgrains	33.0	31.8	26.8	2.8	1.4
Fruits and Vegetables	14.0	13.5	17.9	2.5	6.0
Livestock	20.0	22.7	25.1	4.5	3.8
Fish	2.6	3.0	4.0	3.7	5.0
Total agriculture	100.0	100.0	100.0	3.2	2.9
High-value	32.5	35.9	44.4	4.1	5.0
Rest	67.5	64.1	55.6	2.7	1.5

Source: Gulati, calculations based on data from CSO.

Within crop agriculture, the growth rate of foodgrains has declined from 2.8 per cent in the period of 1980/81 to 1991/92 to 1.4 per cent in the period between 1992/93 to 2002/03 while the growth rate of fruits and vegetables has increased in the same periods from 2.5 per cent to 6 per cent (Table 3.4). The growth rates of high-value agriculture in total have increased from 4.1 per cent to 5 per cent. This compares to declining growth rates in agriculture overall from 3.2 per cent to 2.9 per cent. Joshi *et al.* (2006) decompose the crop income growth over the 1980s and the 1990s into the contribution of yield increases, area expansion, price increases and diversification from low-value crops to higher-value crops. They find that technology/higher yields were the major source of crop growth in the 1980s while rising prices and diversification emerged as the major source of growth in agriculture in the 1990s.

Mittal (2006) projects that the demand for fruits and vegetables in India will increase from 77 million tonnes in 2000 to 183 million tonnes in 2020 assuming an annual GDP growth of 7 per cent over that same period. The increasing demand—and a slow response in supply—shows up in relative prices. Prices in 2006 for fruits and vegetables have increased by 46 per cent and 36 per cent respectively compared to the year 2000. This compares to a 15 per cent increase over the same period for cereals and pulses (Figure 3.1). This makes a switch to these crops attractive for farmers and provides an important new income opportunity for farmers.

Figure 3.1 Evolution Wholesale Price Indices

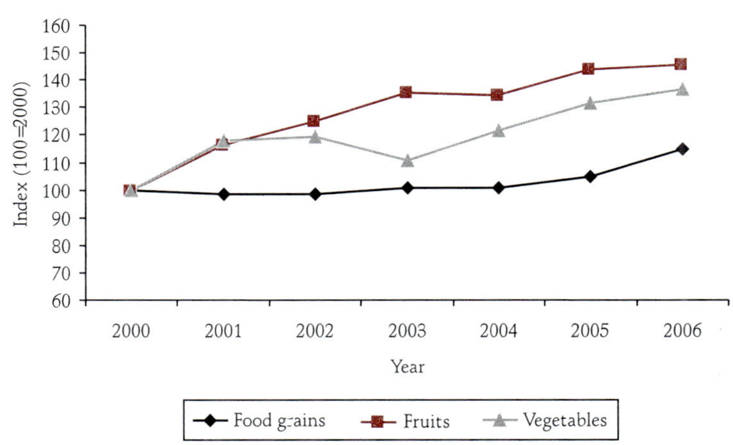

Source: Office of the Economic Advisor, Ministry of Commerce.

We also find significant variation within the fruit and vegetable sector. Table 3.5 below indicates the price variability for major crops over the last six years. It also indicates how consumption per capita of different fruits and vegetables has changed over the last 15 years. We find large variation in prices as well as in consumption. Some vegetables (such as potatoes) and fruits (apples, papaya and guava) have shown significant price increases while others saw significant declines (especially onions). Per capita consumption has been on the increase for most of the fruits and vegetables over the last 15 years but it has been more pronounced for products such as potatoes (+25%), tomatoes (+42%), cabbage (+107%) and bananas (+71%).

Table 3.5 Price and Consumption Trends for Major Fruits and Vegetables in India

	Wholesale price index*			Consumption (g/cap/day)**	
	1998-2000	2001-2003	2004-2006	1990	2005
Vegetables					
Potatoes	100	122	143	35.2	44.1
Sweet potatoes	100	105	102	3.7	2.2
Onions	100	71	67	8.9	10.5
Green peas	100	97	97	5.8	7.2
Tomatoes	100	89	92	12.7	18.0
Cauliflower	100	87	89	10.6	12.4
Brinjal	100	93	89	-	-
Okra	100	96	98	-	-
Cabbage	100	95	102	7.4	15.3
Fresh ginger	100	108	109	0.6	0.6
Fruits					
Banana	100	101	108	16.8	28.7
Mangoes	100	90	94	22.0	21.8
Apples	100	153	197	3.1	3.0
Oranges	100	106	106	5.5	7.2
Cashewnuts	100	99	96	1.3	1.6
Coconut fresh	100	94	87	22.2	21.1
Papaya	100	122	144	1.1	1.6
Grapes	100	101	105	1.2	3.0
Pineapple	100	108	114	2.8	3.1
Guava	100	130	188	-	-

Source: * Office of Economic Advisor, Ministry of Commerce.
** Faostat.

3.2. Modern Food Retail[1]

The organised retail in the food and grocery segment in India is growing fast. Growth is estimated to be significantly higher than 50 per cent.[2] If these growth rates continue, it might not be long—say, by 2015—before the share of organised retail in the food and grocery segments accounts for at least 15-20 per cent; by then it would start to have noticeable impacts not only on unorganised retail in food but all along the food supply chain (Reardon and Gulati, 2008). We discuss briefly the impact of this development on three major stakeholders: consumers, traditional retailers and farmers.

Consumers: Indian consumers spent, on average, about 51 per cent of their total expenditures on food in 2004. The emergence of organised retail gives consumers a wider choice of goods, more convenience and a better shopping environment, among other benefits. This is feasible because organised retail can take several formats, from small neighbourhood stores in densely populated cities, with high real estate prices, to hypermarkets in the periphery where real estate is cheaper. Organised retail can appear small but will likely spread to all local markets, providing the convenience of a neighbourhood store but with procurement on a mass scale that keeps prices

1. This section draws heavily on Reardon and Gulati (2008).
2. The projected growth rates for the next five years are likely to be accelerating.

low and provides greater variety. Moreover, although it is difficult to implement any food safety standards in the traditional retailing environment, modern organised retailers could be thought of an entry point to ensure food safety, not only at the retail end but also all along the supply chain (Reardon and Gulati, 2008).

Retailers: Traditional retailers occupy an overwhelmingly large space in Indian food retail as almost 99 per cent of food and grocery is being sold through traditional retailers. Therefore, what happens to their livelihoods as modern retail expands is a legitimate concern that every policy-maker must recognise. Experience in other countries shows that traditional and modern retail can coexist and grow, albeit at different rates, for many years, usually decades. However, structural changes in retail will surely start affecting large numbers of small retailers at some stage, be it after one or two decades, especially when the overall share of organised retail reaches about 25-30 per cent (Reardon and Gulati, 2008).

Farmers: The experience in other countries reveals that processed food generally occupies the largest share of retail (roughly 65 per cent), followed by semi-processed food (about 20 per cent) and fresh food (about 15 per cent). Although direct links between organised retailers and farmers are possible only for fresh food, many farmers are likely to gain from links to processors, because processors work closely with modern retailers. Processors often work through contract farming. Most studies confirm that contract farmers earn higher profits than non-contract farmers, and this is primarily achieved by lowering marketing and transaction costs and, in some cases, offering better prices. For example, an International Food Policy Research Institute (IFPRI) study of Mother Dairy, Nestle and Venkateshwara Hatcheries showed that contracting is beneficial because it helps farmers cut the cost of cultivation and earn higher profits compared with non-contract farmers (Birthal *et al.*, 2005).

The process of backward integration can be strengthened and expedited if retailers or their specialised procurement agencies not only connect with producers (farmer organisations and processing companies) for all their help but also help them, especially farmers, by providing critical inputs such as technical expertise, extension, finance and insurance, which are scarce or non-existent in the public support system accessed by most farmers. Given the scale on which organised retailers operate, they can bring in banking, insurance and other support systems accessed by most farmers (Reardon and Gulati, 2008).

3.3. Agricultural Export Markets

International trade in agricultural products has been liberalised in recent decades and exchange rates have been adjusted to give greater incentives to exporters, including agricultural exporters (Minot and Hill, 2007). Agricultural markets remain distorted by subsidies in the rich countries and constrained sanitary and phytosanitary barriers, some of which have been raised for protectionist motives. In spite of these barriers, agricultural trade has grown rapidly, particularly horticultural exports from developing countries to developed countries.

Trade in fresh and processed fruits and vegetables is one of the more dynamic segments of agricultural trade. World trade was estimated to have reached US$71.6 billion in 2001, an increase by 30 per cent compared to 1990 (Diop and Jaffee, 2005). Given the high labour requirements in this sector, the low land costs and longer cultivation periods in developing countries as well as the trade incentives given by some developed countries, developing countries have been able to capture an increasing market share of the world trade. While fresh and processed fruit and vegetable products accounted for 17 per cent of total exports from developing countries in 1980-81, this share increased to 22 per cent in 2000-01 and this, despite a significant price decrease over the same period (Diop and Jaffee, 2005). Developing country exports of fruits and vegetables have grown at almost 12 per cent per year over 2000-2005 (Minot and Hill, 2007). Net imports of fruits and vegetables by the United States have tripled since 1990, largely due to imports from Mexico and other Latin American countries. Meanwhile, net fruit and vegetable imports to Europe have more than doubled since 1990 as supplies from Morocco, Egypt, South Africa, Kenya and other countries have expanded. In Asia, Chinese horticultural exports are displacing domestic production in Japan and other countries.

India has overall not been part of these increasing opportunities in export markets. Its exports are a small fraction of both world horticultural exports and domestic production (Mattoo *et al.*, 2007). India's horticultural exports have hovered around 20 per cent of its total agricultural exports. However, they only present a small percentage of domestic production. Mattoo *et al.* (2007) find that onions are the only horticultural product where exports are a significant proportion of domestic production. The only other product where exports are more than 1 per cent of aggregate production is grapes. The most important items among India's horticultural exports are fresh onions, mango pulp and fresh mangoes, dried walnuts and fresh grapes. Dried and preserved vegetables account for a quarter of all horticultural exports. The largest proportion of India's exports of fresh fruits and vegetables goes to countries in South Asia and Middle East while only a small percentage of India's exports goes to industrial countries. In contrast, a large part of India's limited exports of processed fruits and vegetables goes to industrial country markets (Mattoo *et al.*, 2007).

India's exports suffer from three impediments (Mattoo *et al.*, 2007): 1/ high logistical costs because of high delivery costs and logistical inefficiencies; 2/ the high standards required by governments and buyers, especially so in richer countries, and the low standards and weak conformity assessment mechanisms in India; 3/ the trade policy barriers that Indian exporters face in foreign markets including entry prices that discriminate against efficient delivery, tariff quotas that subject imports above specified low levels to harsh tariffs and special safeguards which are a source of considerable uncertainty for successful exporters (Mattoo *et al.*, 2007).

Although food safety and agricultural health standards are designed to reduce the risks of the spread of plant and animal pests and diseases and the incidence of microbial pathogens or contaminants in food, standards can also be used as a trade protection measure and there is increasing evidence that this is the case (Mattoo *et al.*, 2007; World Bank, 2007). Anyhow, these standards are there to stay and meeting increasingly high standards will be a critical challenge in expanding exports to industrial countries.

Agricultural exports also currently make up for only a small part of agricultural high-value production in the state of Uttarakhand. To stimulate exports of high-value agriculture, the state has set up four agri-export zones (AEZ): litchi, floriculture, basmati rice and medicinal and aromatic plants. Three other AEZs are being proposed: walnuts, temperate fruits and off-season vegetables, and organic produce. Both the Central as well as the state government are providing a variety of financial assistance to encourage agricultural exports (APEDA, NHB, Department of Food Processing Industries, SFAC, etc.) as well as fiscal incentives for these AEZ (AEDU, 2006).

It is estimated that Uttarakhand exported 34 MT of litchis in 2004-05 and 67.5 MT in 2005-2006 (AEDU, 2006). In the year 2003-04, more than 3 tonnes of flowers were exported to Kuwait and Dubai (AEDU, 2006). AEDU made an export target of 5 tonnes of flowers in 2006-07. In the year 2005-06, the state had exported 314 tonnes of Basmati rice (AEDU, 2006). AEDU made a target of 400 tonnes in 2006-07. In 2005-06, 27 tonnes of medicinal plants and aromatic plants were exported (AEDU, 2006). AEDU made the target 40 tonnes in 2007. The state is currently also exporting organic produce and culinary herbs and some other high-value crops. However, it is clear that most of the production is still geared towards national markets. When the exports of these crops are approximately valued, we obtain about Rs. 10 lakhs for flowers, Rs. 39 lakhs for basmati rice and about Rs. 80 lakhs for litchi (valued a wholesale price of 1500 Rs./quintal; the median price based on Agmarknet data). The export value for these three crops thus amount to Rs. 130 lakhs (US$ 300,000) or about 0.3 per cent of the annual value of potato production in the state (valued approximately at US$ 100 million).

4 The Current State of High-value Agriculture

4.1. Overview Agriculture

Being a state with diverse agro-climatic endowments, conditions under which agriculture is carried out differ remarkably across areas. Broadly the plains and hills present differing scenarios for agriculture. While commercial agriculture is practiced in the plains, the hill farmers mainly practice subsistence farming. The hills practice mixed cropping, while in the plains in a given season single crops are mostly grown. Irrigated land is freely available in the plains, with over 87 per cent land being irrigated as against a mere 10 per cent in the hills. The seed replacement rate for the plains stands at 15-20 per cent, while for the hills it is 3-4 per cent. Productivity across the same crops also differs greatly between the hills and plains. For instance, the productivity of wheat in the hills is 13.2 quintals/hectare, while in the plains it is 30.45 quintals/hectare. For rice the figures stand at 12.36 quintals/hectare in the hills and 27.49 quintals/hectare in the plains (Uttarakhand State Planning Commission, 2007). This highlights the need for separate approaches for agricultural development in the hills and plains areas of the state.

Another feature typical of hill farming is the small and scattered landholdings. As noted in Chapter 2, the land available for cultivation is a very small share of the total land, due to the large share of forests. Of the total cultivated area, about 50 per cent of landholdings (in number) are sub-marginal, and 21 per cent of landholdings measure between 0.5–1 hectare. 27 per cent of the area under cultivation consists of plots less than 1 hectare in size. Another 26 per cent of land holdings are between 1 and 4 hectares in size, and account for 51 per cent of the total cultivated area. 22 per cent of the cultivated land consists of plots over 4 hectares in size, and these account for 3 per cent of the land holdings in number (ibid.).

Currently, agriculture accounts for Rs. 4,995 crores of the GDP. Within the primary sector, agriculture accounts for nearly 85 per cent, followed distantly by forestry and logging. The contribution of agriculture (including livestock) to GDP during 1999-2000 to 2005-06 ranged between 19 and 27 per cent. Livestock rearing is gradually gaining importance in agriculture as its share in the total agricultural Gross Value of Output (GVO) is rising; from about 28 per cent in 1998 to 30 per cent in 2005-06. Dairying is the main driver within the livestock sector. The fisheries sector which has received some attention in recent years still accounts for only 0.1 per cent of GDP and though its untapped potential, given Uttarakhand's vast climatic and water resources, is important to be considered, currently this sector is on a very small scale in the state, generating a GVO of Rs. 8 crores.

We next turn to crop agriculture. Table 4.1 enlists the area under principal cereals and pulse crops and the annual production for the year 2004-05. In rainfed areas local millets are very important, while in the plains crops such as wheat, rice and maize dominate. There is an emphasis on rice and wheat production in the state in production terms, but the dominance of these crops in terms of area coverage and the percentages of net sown area under them is somewhat lessened. This would be especially true if we looked at the hill areas separately, as the plains tend to dominate in production of wheat and rice.

Table 4.1 **Area and Production of Principal Crops: 2004-05**

Crop	Area in 1000 hectare	% of net sown area	Production (MT)
Paddy	299	39.45	550*
Madua	148	19.53	190
Sawan	72	9.50	90
Maize	30	3.96	44
Kharif pulses	23	3.03	-
Kharif oilseeds	24	3.17	-
Wheat	394	51.98	793
Barley	24	3.17	33
Rabi pulses	22	2.90	-
Rabi oilseeds	17	2.24	-
Total pulses	-	-	28
Total oilseeds	-	-	35

Note: * Production statistics available for rice.
Source: State Planning Commission, Government of Uttarakhand.

Uttarakhand has much potential for agricultural diversification and development. We compare the shares of different groups of crops in the Gross Value of Output (GVO) for agriculture, in order to draw insights into the relative importance of crop groups. At the all-India level, cereals account for 32 per cent of the agricultural GVO, which is very close to Uttarakhand's 34 per cent. Given that most of the endemic cereals are produced on a small scale and often for subsistence needs, it is hoped that over the years this share may marginally come down, making way for higher value crop groups.

Pulses are an important component of the Indian diet and account for 4.8 per cent of the agricultural GVO nationally. They however command lesser importance in Uttarakhand, accounting for only 2.4 per cent, half the national average. In contrast, pulses are a very important component of the agricultural economy of states such as Madhya Pradesh, Maharashtra and Andhra Pradesh. Sugar accounts for a huge 22 per cent of the agricultural GVO of the state. This makes Uttarakhand the state with the highest share of sugar in total agricultural output, far in excess of the national average of 8.3 per cent. This represents however a small absolute value of sugar when compared with the much larger states of Uttar Pradesh, Karnataka and Maharashtra. Even in these states, the share of sugar is capped at 20 per cent for UP, and roughly 14 per cent for Karnataka and Maharashtra.

Within high-value crop agriculture, Uttarakhand produces an estimated 345 thousand tonnes of fruits and 493 thousand tonnes of vegetables every year (GoUA, n.d., a). Presently, 39 per cent of the cultivable land in the state is under horticulture crops. The rate of increase in coverage of fruit crops has been 4.2 per cent, of vegetables 7.2 per cent, flowers 17.7 per cent and spices 9.7 per cent from the period 2002-03 to 2005-06 (GoUA n.d., b). With market development, as horticultural cultivation becomes more lucrative, it is expected that these rates will remain high and that horticulture will account for larger shares in agricultural output and value.

Currently fruits and vegetables account for 27.2 per cent of the agricultural GVO, less than a percentage point higher than the national average of 26.7 per cent. The situation appears low when compared with the other hill states of Himachal Pradesh and Jammu and Kashmir where fruits and vegetables make up roughly 62 and 58 per cent of the agricultural GVO, respectively. Table 4.2 shows the importance of the different fruits and vegetables grown in the state.

Agro-processing overall is not well developed and it is estimated that it accounts for a mere 1.5 per cent of total fruits and vegetable production in the state. Recognising the important role that the processing of horticultural products can play in the state, the government of Uttrakahand, as well as the Centre has provided a number of incentives and support schemes for the growth and development of agro-processing in the state. Support for this sector is an important component of the State's Industrial Policy, and currently a subsidy programme funded by the Central Ministry of Food Processing Industries (MoFPI), and administered by the Directorate of Horticulture in the state government is in place. The APEDA and the National Horticultural Board also provide subsidies to processors in the state. Till the year 2004-05, 19 horticulture-based processing units that came up in the state were subsidised to the tune of Rs. 7214 lakhs. 16 cereals, milk and other commodity-based processing industries received Rs. 9158 lakhs in financial assistance from the state. The units set up are detailed in Table 4.3.

Table 4.3 shows that the agro-processing industry is heavily concentrated in the district of Udham Singh Nagar, and that the hill districts have received virtually no investment (a mere 4 units between Nainital and Almora). Poor roads and connectivity presumably pose onerous challenges to the spread of agro-processing industries in the interiors. Also there is a marked emphasis on horticulture-based processing (typically pulp, jam, squash, juice and pickle-making units) and the potential of consumer food-based industry seems to be sub-optimally tapped.

Table 4.2 Area and Production of Horticultural Crops

	Year 2005-06		Year 2006-07 - Tentative	
	Area hectares	Production M.T.	Area hectares	Production M.T.
Fruits				
Apple	28,800	112,320	29,725	117,414
Pear	14,250	99,631	14,540	101,780
Peach	6,770	36,532	8,166	44,505
Plum	9,198	39,775	9,400	40,890
Apricot	9,054	30,375	9,100	30,576
Walnut	18,590	16,324	18,838	16,766
Citrus	26,142	122,934	27,900	132,525
Mango	36,020	106,630	36,900	109,962
Litchi	7,941	13,618	8,980	15,535
Other	24,980	114,311	25,600	117,760
Total fruits	*181,745*	*692,450*	*189,149*	*727,713*
Vegetables				
Pea	9,645	60,932	10,610	67,579
Radish	3,621	45,059	3,983	49,709
French beans	4,107	29,479	4,518	32,573
Cabbage	4,036	50,192	4,440	55,317
Cauliflower	1,950	30,847	2,145	33,998
Onion	3,138	31,731	3,452	34,932
Capsicum	1,909	10,600	2,100	11,675
Okra	2,682	18,459	2,950	20,326
Tomato	7,967	76,993	8,764	85,183
Bringal	1,595	20,612	1,754	22,686
Other veg.	10,073	86,169	11,080	95,069
Total veg.	*50,723*	*461,073*	*55,796*	*509,047*
Potato	22,157	450,637	24,373	495,984
Flowers	562	641	619	710
Spices	6,626	49,903	7,287	54,986

Source: Directorate of Horticulture.

4.2. Opportunities for High-value Crop Development

The state of Uttarakhand is characterised by a significant number of opportunities in high-value agriculture. They include most importantly the presence of a high number of endemic crops, diversity in agro-climatic conditions, possibilities to produce for 'off-season' markets, organic production practices, the relative high education of producers, a strong agricultural research capacity, an active civil society, a competitive production environment and a location relatively close to terminal consumer markets, at least for part of the state (the plains and some of the mid-mountains).

Due to its diversity (linked to a large variation in altitude, ranging from about 300 metres to 7,800 metres), the state of Uttarakhand is able to produce a large variety of high-value crops, ranging from different types of vegetables to temperate fruits such as apple, peach, pear, plum, walnut and tropical fruits such as mango, litchi, papaya and guava. The state of Uttarakhand is further characterised by a large number of endemic crops. These crops have large biodiversity

Table 4.3 Location and Category of Agro-processing Units Supported by the State[1]

	Horticulture-based	Cereal-based	Milk based	Consumer food based	Oil based	District totals
Haridwar	1	1	1			3
Dehradun	3					3
Udham Singh Nagar	7	7	2	3	2	21
Nainital	3					3
Almora	1					1
Total	15	8	3	3	2	

Source: Horticulture Department, Uttarakhand. Printed in GoUA (n.d.) 'c'.

value and might also have large untapped marketing potential. Such crops exist within the group of medicinal and herbal plants and also as grains and pulses. Crops such as black soybean and finger millets (*madua*) are almost exclusively grown in the state of Uttarakhand and might have large untapped marketing potential.[2]

The hill producers in the state of Uttarakhand have the advantage to tap into 'off-seasonal' markets in India, meaning that they are able to supply fresh produce to large terminal markets such as New Delhi when few other states are able to do so. These off-seasonal products concern most of the fresh vegetables as well as some fruits. The only other states that have similar production conditions are the states of Himachal Pradesh and Jammu and Kashmir.

A large number of crops in the hills of Uttarakhand are grown organically. Due to the lack of irrigation, the higher costs of inputs linked with its remoteness as well as the higher production risks (due to natural calamities), modern inputs are used little in the hills. While the lack of modern input use leads on average to lower productivity, the output has however the potential advantage for being sold in the rapidly increasing organic premium market. As farmers are used to organic production practices, the conversion to these production practices would entail little costs and behavioural change. Combined with the Himalayan brand image, this might allow access to some promising and emerging markets.

Socio-economic characteristics are also a large advantage for rural areas in Uttarakhand. Education levels are relatively high, especially compared to other neighbouring states. Civil society is strongly represented as there are a large number of NGOs that are registered. The state is further home to the renowned G.B. Pant University of Agriculture & Technology and has multiple research centres in different districts spread over the state.

Labour and land prices are relatively low which makes it a competitive state for agricultural entrepreneurs to invest in. Mobile phones and electricity have recently become widely available in rural areas making it easier to do business. Finally, customers (from terminal markets such as Delhi, rapidly increasing demand from rural towns in the state of Uttarakhand as well as the increasing demand from tourists, pilgrims as well as others) are close by and are demanding more—as well as better quality—produce. This implies that an important 'pull-factor' might lead to positive incentives for agricultural improvements.

1. Includes data on 26 completed and 5 pipeline projects.
2. A crop such as finger millet, which is widely grown in the traditional environments, has high nutritional value and is for example used in fortified food supplements.

4.3. Challenges for High-value Crop Development

Agriculture in Uttarakhand faces also significant challenges that limit the competitiveness of its farmers with farmers in other Indian states and outside India. These include the high number of small scattered farms creating problems of aggregation and transport costs, migration and land conversion, increasing water and climatic change problems, environmental vulnerability, wildlife attacks, and a problematic regulatory environment.

While the plains have a large number of relatively modern farms in densely farmed areas, the hills on the other hand are characterised by small scale farming where farms are scattered, commercial agricultural surplus is low and aggregation and transportation costs are high. In such an environment, transaction costs are high and remunerative producer prices low. Box 4.1 shows how such an environment might make it unprofitable for new agricultural enterprises to invest. Any agricultural development strategy would have to take into consideration these high transportation costs in the mid-mountains but especially so for the high-mountains as the high transportation costs and special ecological conditions limit the choices of products that can be marketed.

Box 4.1 The Costs of Doing Commercial Agriculture with Scattered Hill Farmers in Uttarakhand

The IndoDutch Horticulture Technologies Pvt. Ltd. started commercial agricultural activities in Uttarakhand (Bhimtal) last year. The firm focuses on growing flower bulbs for the local market and luxury fruit (such as blueberry, raspberry, and strawberry) mostly for Delhi outlets and new plantings of apple trees to be sold locally in Uttarakhand as well as in Kashmir. The large advantage for the firm in the hills of Uttarakhand is the climate (the long growing period as well as the off-seasonal advantage). The big disadvantage is the bad road infrastructure and the costs of dealing with scattered farms. This creates all kind of costs. First, the firm deals with a number of scattered farmers located 1 to 2 kms from the main road which creates costs doing the extension work for the introduction of these new technologies and carrying the produce to the main road. Second, in this type of environment, it takes 20 labourers two weeks to plant 75,000 trees on 3 ha of land. This work could normally be done by 3 persons in 4 days on flat land. Third, in the hills, it takes 5 people 3 days to do the spraying. On flat land, spraying can be done in 4 hours (by a tractor). Fourth, 2 people are continuously employed to ensure the right working of the drip irrigation system. In flat land, this could be done "with a push of a button". Given the costs of dealing with scattered farmers, the firm intends to diversify towards Jammu & Kashmir where flat land is more easily available.

The relatively high education level of the people in Uttarakhand and the liberalised economic environment in India has triggered fast economic growth, increasing urbanisation and growth of the service sector, migration of agricultural workers and farmers to off-farm activities, and increasing labour shortage in hill agriculture. This has led to a reduction of yields in upland agriculture and even the abandonment of agricultural land in the hills. In the plains, the fast economic growth and urban spread seems to increasingly lead towards the conversion of agricultural land to non-agricultural uses.

While exact data are difficult to obtain, water problems and climatic change seem to have dramatic impacts on traditional production practices and yields of agricultural products in some districts of the state (for example, see *Down to Earth* (2006) for a description of the water problems in the district of Almora). Access to water (especially potable water) is often mentioned by hill farmers as the major constraint in their daily life. Compared to decades ago, some hill areas have seen the drying up of a number of water sources and access to drinking and irrigation water has become problematic, even though the increasing adoption of rainwater harvesting has partly alleviated this problem. Farmers also state that climatic changes have led to the more unequal spread of rainfall over the year and the majority of the rainfall has now become more concentrated in a couple of months during the monsoon period. This affects agricultural activities adversely.

The production of off-seasonal vegetables has in large parts of the state of Uttarakhand had a big beneficial outcome on small farmers. It is a success story that has improved the lives of a significant number of poor farmers in the hills. However, the production of off-seasonal vegetables has often gone hand in hand with the seemingly excessive use of chemicals—often subsidised by the government—which are having a bad effect on the environment through its leaching into water resources. This excessive use of chemicals might not only be unsustainable for the environment but might also lead to negative effects on food safety.

Wildlife attacks are further increasingly mentioned as a major constraint to the production of high-value agricultural crops in the state. Especially attacks by wild boars seem to lead to significant losses in agricultural production. While exact statistics are lacking, the number of these wild boars is stated to have exploded in the last 15 years. Current legislation does not allow wild boars to be shot and their population explosion emerges as a major risk factor in agricultural development for large parts of the state.

The last challenge for the development of high-value agriculture in Uttarakhand—but arguably also one of the most important—is the lack of an appropriate effective legal framework for agricultural marketing. The state of Uttarakhand is one of the few states in India that still adheres to a non-amended Agricultural Produce Marketing Committee (APMC) Act. Under this Act, agricultural marketing is highly regulated by the state government and although well-intended, it creates a marketing environment—as stated by different stakeholders—that is characterised by high marketing and transaction costs due to licensing rents, corruption, lack of transparency and lack of competition between traders. For example, while the regulations of the APMC Act state that market taxes should be 2.5 per cent and commission rates 3 per cent and that these rates are only charged to the buyer, the reality in the field paints however a completely different picture (Box 4.2). This is analysed in greater depth in the next chapter.

Box 4.2 The Costs of Marketing in the Local *Mandi* System

CHIRAG (the Central Himalayan Rural Action Group) conducted a study of the marketing costs of supplying the local *mandi* of Haldwani by farmers located around Ramgarh (two hours from Haldwani). Payment slips handed out by commission agents to hundreds of farmers were collected as to estimate the different costs to supply produce to the local market. The overall costs of sending produce to the local market ranged between 9.9 per cent of the total value for peas and 24.2 per cent for pears. The costs were split between commission rates (4 per cent for potatoes and 7 per cent for all other produce), 'sampling costs' (1 per cent), transport costs (varying between 1.5 per cent and 10.6 per cent), labour charges for loading and unloading (varying between 0.24 per cent and 1 per cent) and postage and communication (varying between 0.1 per cent and 1.82 per cent). These costs are the monetary costs. They do not include the costs of the farmer to bring produce to the roadside or the time lost for doing the transaction in the Haldwani market (travelling back and forth and waiting until the transaction is finished might take up a day or even longer).

5 High-value Crop Marketing

5.1. Introduction

In this chapter, we turn attention to some of the key policy and institutional factors affecting the marketing of high-value crops in Uttarakhand today. First, we look at the legal issues concerning marketing of agricultural produce in the state. We discuss the details of the regulated market system as stated in the APMC Act. Second, we present an overview of effective practices on these markets by relying on a primary survey that was conducted at the end of 2007 on the two largest wholesale markets in the state, the Haldwani and Dehradun wholesale market.

5.2. The APMC Act

Under the current policy regime in the state, the bulk of trade in agricultural commodities takes place at the wholesale *mandis*, run and operated by the Uttarakhand Krishi Utpandan Mandi Parishad (Agricultural Produce Marketing Committee) which oversees and governs the functioning of the 20 *mandis* that have been established across the state. *Mandis* are essentially designated market spaces run by the government for wholesale trade in agricultural commodities between farmers and private wholesale traders, as well as between wholesalers and retailers. These *mandis* also serve as sites for inter-state trade.

The *mandi* premises are known as the 'market yard'. This must be distinguished from the '*mandi* area', which is the entire territory under the purview of a particular *mandi*. This means that all the *mandi* tax chargeable to the agricultural output produced in the area must be paid to that particular *mandi*, even if it is not physically traded within its premises. The market yard comprises shops of typically four kinds which can be leased out to private commission agents (CA), locally known as *aadhtis* who hold valid licenses to operate business in the *mandi*. Commission agents are private persons who organise auctions in their shops on behalf of farmers to sell the produce to the highest bidding traders. The intended aim of the commission agent-mediated system has been to help farmers get the best price for their produce as the system is based on competitive bidding, and the farmer can directly witness the auction of his produce and ascertain fairness (e.g. Shiva, 2007; Acharya, 2004).

The *Mandi* Committee allots shops to desirous agents who meet basic eligibility criteria (based on nationality, solvency and other not particularly restrictive criteria) upon payment of a license fee. Based on availability and license fee paid, different categories of shops are allotted. The most expensive shops are the A category ones which usually have an extended trading and storage in addition to a separate space for a small office. B category shops have smaller trading areas and a much smaller separate sitting area, instead of a specific office. C category shops have even smaller trading areas and no designated sitting area. Finally, small tin-sheds in a large trading hall may be allotted to a lower category of applicants to conduct auctions in the common premises but still have a designated area where traders and farmers may identify and transact with a particular commission agent. Note that this description may vary slightly across *mandis*.

Typically, the number of licence holders is far in excess of the number of shops available in the *mandi* yard. Legally, such licence holders have equal right to transact business, but effectively the lack of space for trading volumes of say, fruits and vegetables severely curtails their business. It is estimated that in Dehradun *mandi*, only 5 per cent of licence-holding commission agents do not have a shop. Licences once awarded can be easily renewed annually. Invariably members of the same family transact business at the same shop passed down to posterity. Since the number of shops is practically fixed and licences are rarely returned by holders, the existing advantage that licence holders have in terms of having secured the shop at a given rate is only fortified over time.

Table 5.1 Taxes and Other Charges: APMC Uttarakhand

Heading	To be paid by	Rate
Offloading expenses	Farmer	Re. 0.50 per quintal
Sorting and stacking	Farmer	Re. 1.00 per quintal
Commission charge	Trader	1.5% for cereals and pulses 3% for fruits & vegetables
Mandi tax[1]	Trader	2%
Development cess	Trader	0.5%
Brokerage	Trader	0.5%
Weighing expenses	Trader	Re. 0.50 per quintal
Labor charges (filling sacks and transporting them)	Trader	Re. 1.00 per quintal

Source: Pamphlet printed by APMC Haldwani.

Table 5.2 Contributions to be made to *Mandi* Directorate, Rudrapur by Different *Mandis*

Annual income generated by *mandi* taxes and other income	Share of *Mandi* Directorate
> 50 lakhs	50% on the first 50 lakhs, 75% for every rupee thereafter
40–50 lakhs	50%
30–40 lakhs	40%
20–30 lakhs	30%
10–20 lakhs	20%
5–10 lakhs	15%
< 5 lakhs	10%

Source: Uttarakhand Krishi Utpadan Mandi Parishad, Rudrapur.

The rates of commission and other charges have also been fixed by the APMC, and these are widely announced in the *mandi* premises. However, in reality the prevailing rates are very different from the stated and fixed rates in the APMC (see Section 3). Table 5.1 gives the prescribed legal rates of commission, tax and other expenses that can be charged at the *mandis*.

The state of Uttarakhand has 20 *mandis* spread across the state, with the apex Uttarakhand Krishi Utpadan Mandi Parishad headquartered in Rudrapur. For the state as a whole, about 64 per cent of the income earned by the individual APMCs is from *mandi* taxes. Other incomes and the development cess account for the remaining 19 and 17 per cent respectively. The development cess is charged by the state government and is not utilised by the *Mandi* Directorate at all. It is sent directly to the state government for its various development schemes. The *mandi* taxes and other income however are to be utilised by the APMC, and a certain contribution is expected to be made to the *Mandi* Directorate. The contributions are made according to the model depicted in Table 5.2.

The *Mandi* Directorate can use the funds thus obtained for meeting its

1. Items on which *mandi* tax has been levied at one Uttarakhand *mandi* need not do so at any other *mandi* in the state. For arrivals from Uttar Pradesh, the rate of tax is 1 per cent.

administrative expenses and for a number of development activities it may choose to carry out in the state. These may particularly be undertaken to help develop infrastructure and support in some of the poorer/weaker *mandi* yards, and also for development activities in general, including the construction of roads, provision of scholarships, etc.

The money remaining with the individual APMCs may be utilised as follows: 1/ No more than 20 per cent of the revenue may be spent on administrative costs including costs of employees' salary, upkeep and maintenance of inventories etc. Of the remaining amount (80 per cent), three-fifths is to be spent on the *mandi* yard, and would include building and upkeep of shops, repairs and maintenance of roads in the *mandi* premises, cleanliness and provision of sanitation services etc., in the *mandi* premises. The remaining two-fifths may be spent on development activities in the *mandi* area; 2/ Each APMC Managing Committee drafts a proposal by passing resolutions that articulate priority areas that merit investment and expenditure and sends this to the APMC Board for approval. After it is approved, the proposed projects may be undertaken. For proposals where the expenditure/investment required is less than Rs. 5 lakhs, no such approval is necessary.

The Model Agriculture Produce Marketing Commission (APMC) Act evolved by the Central government, and circulated among the states in order for them to amend the existing acts and bring them in line with some of the more far-reaching aims envisaged in the model Act is critical for the rapid transformation and modernisation of agriculture in Uttarakhand today. However, unlike most other states, Uttarakhand is one of the last few states not to have amended the prevailing act in accordance with the model Act which seeks to pave the way for greater private sector participation in agriculture, opening up the possibility of contract farming and enhancing the farmers' options of agents he could potentially sell to. Currently the UP Krishi Utpadan Mandi Adhiniyam, 1964 is in vogue in the state, though a draft bill (dated 2005) is also in circulation, which it is believed may be introduced in the Uttarakhand Assembly shortly.

5.3. Results from a Wholesale Market Survey

5.3.1. Data and Methodology

The nodal point in horticultural marketing in Uttarakhand is the wholesale market and the majority of marketed produce passes through it (Fafchamps *et al.*, 2007). The major aim of the survey that was conducted was to better understand the activities on these wholesale markets for fruits and vegetables. After initial assessments it was felt that wholesalers and brokers were not reliable sources of information as their statements were in strong contrast to statements by the persons whom they interacted with. It was thus decided to interview buyers and sellers that just completed a transaction on these wholesale markets and piece together the functioning of brokers and wholesalers based on the interviewees' declarations.

A formal survey was conducted in December 2007 on the two main wholesale markets of Uttarakhand, the wholesale market of Dehradun and the wholesale market of Haldwani. Dehradun, the state capital, is the most important market for the Garhwal hills and the plains areas around it. The chief crops traded here include tomato, cauliflower, bottle gourd, radish and green peas among vegetables and a large number of fruits. The Dehradun market in the year 2007 had 13 category A shops, 34 category B shops, 78 category C shops, 56 category D shops, 10 category E shops and a number of traders and commission agents operating out of tin-sheds.

The Haldwani wholesale market is the main centre of trade for fruit and vegetable produce from the Kumaon hills as well as much of the Terai plains. The main crops traded here are potato, cauliflower, French beans, capsicum, green peas and tomato among vegetables, and apple, peach, apricot, plum and mango. The market has about 230 designated shops of A, B and C categories given out to licence-holding commission agents, and another 20-odd commission agents operate out of tin-sheds (About 250 commission agent licences have been given out).

The survey focused on the vegetables, cauliflower and green peas. At the time of the survey, these were the two vegetables that were being marketed in large amounts in the areas around Haldwani and Dehradun. A total of 480 surveys were conducted of which 240 on the Haldwani wholesale market and 240 on the Dehradun wholesale *mandi*. Half of the surveys were conducted with farmers and half of them with retailers. Half of the agents are involved in green peas and half of them in cauliflower. Farmers and traders were both randomly selected. Farmers were interviewed on the wholesale market while retailers were interviewed on the major retail markets of the city.[2]

The survey contained detailed questions on demographic background of the interviewees, the reasons for the choice of the marketing channel and the broker, and on linkages with the broker used in the last transaction of cauliflower or green peas. Then, information was asked on the last transaction, including prices and costs, quality characteristics of the product, quality and quantity assessments by buyers, and the transactions costs that were faced during the last transaction. The survey finished with questions on wholesale market practices in general.

5.3.2. Descriptive Statistics

We start with descriptive statistics on the farmers and retailers who participated in the survey (Table 5.3). There is little difference with respect to demographics between farmers and retailers. While farmers are slightly older (47 years *versus* 37 years for retailers), the level of education and the size of the households are similar. The majority of the farmers and retailers are Hindu (about three quarters) and about 40 per cent of them are member of a scheduled caste, tribe or other backward caste. Farmers seem slightly poorer: 29 per cent of them carry a BPL (Below the Poverty Line) and 65 per cent an APL (Above the Poverty Line) card.[3] This compares to 30 per cent and 51 per cent respectively for retailers. On the other hand, while 47 per cent of the farmers own a mobile phone, this is only as high as 25 per cent for the retailers.

The majority of farmers and retailers depend on other agricultural products in their business. 83 per cent of the farmers sold other products over the year and 85 per cent of the retailers sold another product over the last two weeks. The two products do however have a large importance for these agents: It makes up 75 per cent of the annual monetary income of farmers and represents 41 per cent of the turnover of retailers over the last two weeks. Both have similar experiences in dealing with the product under study. As could be expected, farmers and retailers have different visiting habits on the wholesale markets. Retailers visit almost every day while farmers come on average 23 times a year. Few farmers (16 per cent) and retailers (2 per cent) visit other markets.

After this brief introduction on the set-up and the characteristics of our sample, we now turn to the question of what is seemingly broken in the current marketing system. We will discuss, consecutively, the effect of regulations, the role of brokers and wholesalers, information transmission, service delivery and competition.

---

2. The three of four major retail markets were selected in every city. Twenty or less retailers of each crop were interviewed at each of these retail markets. They could not be interviewed at the wholesale market as they were often under time pressure to leave the wholesale markets as to start their retail activities

3. Distributed to poorer households by the government as to allow them cheaper access to basic necessities.

Table 5.3 Descriptive Statistics

		Unit	Farmers		Retailers	
			Avg or %	St. Dev.	Avg or %	St. Dev.
Demographics						
Age		years	47.2	10.4	37.3	9.8
Level of education		years	5.3	4.1	4.8	3.4
Household size		number	8.0	3.0	7.0	2.3
Member of scheduled or backward caste/tribe			%	37		39
Wealth						
Has a BPL card		%	29		30	
Has an APL card		%	65		51	
Own mobile phone		%	47		25	
Land owned		*begha* (≈ 1/15 ha)	19.0	23.0		
Own tractor		%	29			
Own cattle		%	88			
Product characteristics (cauliflower, green peas)						
Average sales	both products	kg per day			36.2	31.7
	cauliflower	kg per day			28.3	16.1
	green peas	kg per day			44.2	40.3
Average production	both products	tonnes per season	15.4	18.1		
	cauliflower	tonnes per season	25.4	18.9		
	green peas	tonnes per season	5.5	10.0		
Sell other products	both products	% yes			85	
Importance in monetary income	both products	avg %	75.2	23.9	41.3	30.3
Experience with…	both products	years	12.6	10.9	11.6	8.3
Land cultivated of…	both products	*begha* (≈ 1/15 ha)	21.5	23.5		
	cauliflower	*begha* (≈ 1/15 ha)	23.1	20.8		
	green peas	*begha* (≈ 1/15 ha)	20.0	25.9		
Marketing behaviour						
Distance to wholesale market		km	42.1	34.8	3.5	2.1
Visits on this market		visits last 2 weeks			10.0	3.4
		visits this year	23.5	18.7		
Time spent on the market		hours	3.5	2.2	2.3	1.0
Visit of other *mandi*		%	16		2	

5.3.3. Problems with the Broker System

5.3.3.1. Ineffective Regulations

An important rule of the APMC Act in vogue in Uttarakhand states that the broker rates are 3 per cent and that 2.5 per cent tax on each transaction is to be paid to the market officials. Both these charges are stated not to be paid by the farmer. We test with the data that were collected with farmers and retailers to what extent these rules are respected. We do so by asking about the

Table 5.4 Characteristics of the Most Recent Transaction of Farmers and Retailers

	Unit	Farmers		Retailers	
		Avg or %	St. Dev.	Avg or %	St. Dev.
Values and quantities					
Cauliflower					
Quantity traded - mean	kg	915	579	56	55
Quantity traded - median	kg	800		40	
Value transaction - mean	Rs	4241	2793	282	250
Value transaction - median	Rs	1400		200	
Green peas					
Quantity traded - mean	kg	220	190	52	40
Quantity traded - median	kg	150		50	
Value transaction - mean	Rs	1933	1549	514	461
Value transaction - median	Rs	3600		355	
Commission rates charged					
average	%	5.8	2.8	2.6	3.0
median	%	7.0		0.0	
Number of brokers that charge...					
0%	%	15.4		56.7	
4%	%	7.9		0.0	
5% - 5.5%	%	8.3		1.2	
6% -6.5%	%	18.3		41.2	
8%	%	49.6		0.4	
10%	%	0.4		0.4	
Costs on the wholesale market					
Commission rates	% of gross price	5.8	2.8	2.6	3.0
Labour costs	% of gross price	1.3	1.6	0.5	0.7
Phone costs	% of gross price	0.0	0.0	0.0	0.0
Weighing costs	% of gross price	0.0	0.0	0.0	0.0
Mandi taxes	% of gross price	0.0	0.0	1.4	1.7
Sampling costs	% of gross price	0.3	0.3	0.0	0.0
Total	% of gross price	7.4	3.8	4.6	2.4

costs that farmers and retailers faced during their last transaction on the wholesale market as well as by looking at the effective net prices realised by farmers and retailers. It is noteworthy to first look at the size of these transactions (Table 5.4). The average size of these transactions are small, for the retailer around 50 kgs of produce and representing between US$5 and US$15, depending on the crop. They are significantly larger for farmers indicating that brokers split up the lots of the farmers.

Table 5.4 further presents the different broker rates paid by farmers and retailers to brokers.[4] We find that the average broker rate paid by farmers and retailers is 6 per cent and 3 per cent

4. While taxes are in principle paid by the brokers, it is unclear if they effectively do so for every transaction. In Haldwani, farmers receive a slip when they enter the market on which they indicate how much they bring in and with which broker they will deal with. This gives the broker seemingly less room for maneuvering.

respectively. 50 per cent of the farmers pay 8 per cent while 15 per cent state to pay 0 per cent (Table 5.4). Broker rates differ significantly from the prescribed ones and regulations by the APMC are not respected in two ways. First, broker rates are charged to farmers, in contrast with regulations.[5] Second, the combined rates of farmers and retailers are higher than the prescribed 5.5 per cent. It is also interesting to note the different practices between the different wholesale markets. While most of the brokers charge only the farmers in the Haldwani market, farmers and retailers are charged alike (but thus higher rates in total) in the Dehradun market. It is unclear why this rate setting evolved differently for these two markets, despite being governed by the same APMC laws.

Table 5.4 further documents the other costs that farmers and retailers face on the wholesale market. Labour costs (for loading and off-loading) represent 1.8 per cent of the total value of the transaction while phone and weighing costs are negligible. A number of farmers also face 'sampling costs' representing 0.3 per cent of the value of the transaction. This broker practice involves brokers to take a head of cauliflower or a quarter kg of peas and set it apart for themselves. The total reported costs faced by the farmers (7.4 per cent) and by the retailer (4.6 per cent) on the wholesale market amount to 12 per cent of the value of the lot.

While statements on costs are informative, the real benchmarks are the net payments that farmers receive and that retailers pay when they leave the wholesale market. Figures 5.1 and 5.2 show the net prices that farmers received and that retailers paid for. First, they illustrate the large price variation for these products over the time of survey, often due to location and quality differences. Second, they show the clear parallel leftward shift of the net price of the farmer compared to the price paid by the retailer.

Figure 5.1 Cauliflower Prices

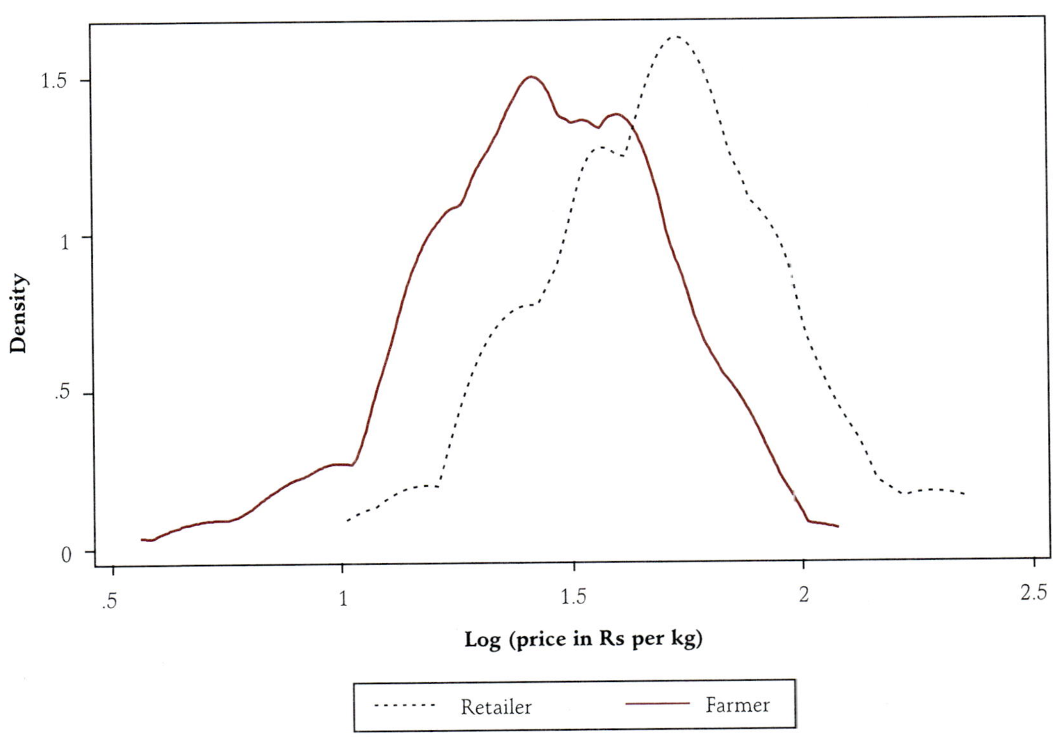

5. We do not discuss the sense of this regulation. One would expect in competitive markets that it should not matter for price formation at the farm level to whom the broker rates would be charged.

Figure 5.2 Green Pea Prices

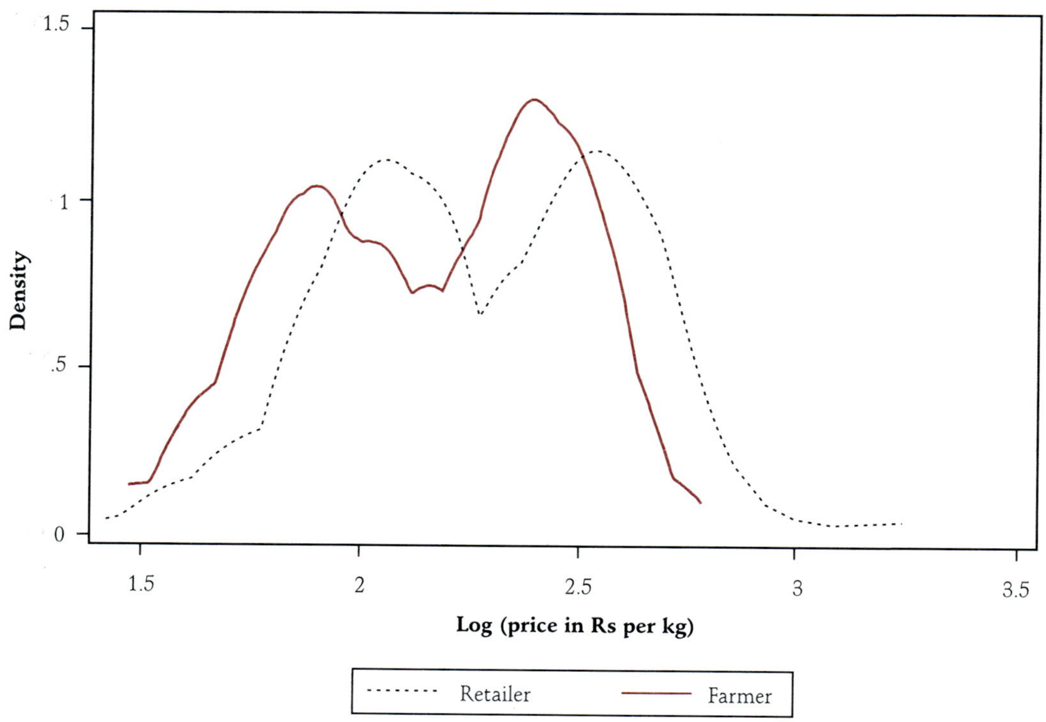

To formally test for the size of the differences in these net payments, we run a regression where we link the log of the price per kilogram paid by the buyer and received by the seller on a dummy of the retailer and the location of the market. The results are shown in Table 5.5. They indicate that the price that the retailers pay is significantly higher than the price received by the farmer for both cauliflower and green peas. The difference is as high as 13 per cent in the case of green peas and 26 per cent in the case of cauliflower. As price differences could be explained by other potential determinants such as the quality of the product as well as the day of the transaction, we add these additional controls in the regression (bottom Table 5.5). The coefficients stay largely significant and the size of the coefficient is robust. For both specifications we use a formal F-test to verify that the price difference between retailer and producer is higher than the prescribed 5.5 per cent. The results indicate that the margin is significantly higher than the prescribed rates and that marketing regulations on margins are thus not respected.

5.3.3.2. The Confusing Role of Wholesalers *versus* Brokers

Two different licences are given out by the APMC market officials on the wholesale market, i.e., a broker and a trader licence.[6] While a broker licence allows for the organisation of an auction and for the facilitation of sellers and buyers to find each other, a wholesaler on the other hand becomes the owner of agricultural produce and is allowed to sell the produce, but not necessarily through an auction. We find in our surveys that the same persons hold often both licences for the same products. While 95 per cent of the farmers state that they sold their produce through a formal auction, in practice this might be less (Table 5.6) as a quarter of the farmers state that the

6. The marketing system is characterised by more licences. Anybody involved in food and agricultural trade (except farmers) need a licence of the APMC.

Table 5.5 **Determinants of Vegetable Prices (dep. var=log(price per kg))**

		Cauliflower		Green peas	
	Unit	Coefficient	t-value	Coefficient	t-value
Parsimonious specification					
retailer	yes=1	0.263	9.23	0.127	4.12
Dehradun market	yes=1	0.282	9.91	-0.405	-13.10
intercept		1.296	52.47	2.345	87.54
Number of observations			240		240
F(2, 237)		91.69		94.25	
Prob > F		0		0	
R-squared		0.436		0.443	
Adj R-square		0.432		0.438	
F-test:					
Price difference between farmer and retailer is higher than prescribed 5.5%					
F(1,239)		53.37		5.48	
Prob>F		0.00		0.02	
Including controls for quality and the day of transaction					
retailer		0.244	3.37	0.128	4.53
medium size	yes=1	-0.128	-3.68		
small size	yes=1	-0.272	-4.67		
mixed size	yes=1	-0.119	-3.13		
number of peas per shell	number			-0.023	-1.48
presence of spots	1=yes	0.067	1.90	-0.015	-0.37
rotten material	1=yes	0.054	1.37	-0.128	-3.05
less bright color	1=yes	0.029	0.68	-0.098	-2.41
Dehradun market		0.308	7.99	-0.469	-14.11
		day of transaction included but not reported			
Intercept		1.713	12.39	2.506	20.05
Number of observations		239		236	
F(x, 220)		19.38		24.15	
Prob > F		0		0	
R-squared		0.613		0.622	
Adj R-square		0.582		0.596	
F-test:					
Price difference between farmer and retailer is higher than prescribed 5.5%					
F(1,239)		42.06		6.66	
Prob>F		0.00		0.01	

buyer was the broker. Two-thirds of the retailers state that the broker (and not the farmer) was the seller.

Table 5.6 Role of Wholesalers and Brokers

	Unit	Farmers		Retailers	
		Avg or %	St. Dev.	Avg or %	St. Dev.
Transactions through auctions	%	95		88	
Number of buyers in auction	Number	7.7	5.5	6.5	3.3
Type of typer					
The broker	%	26			
Retailer in town	%	29			
Reseller on the *mandi*	%	18			
Trader outside town	%	19			
Don't know	%	9			
Type of seller					
The broker	%			65	
Farmer	%			25	
Trader	%			9	
Other	%			1	

Confusion on the exact role of brokers might lead to an conflict of interest. As a broker is paid through a percentage of the final price, he has an incentive to find a high price for the seller. A wholesale trader on the other hand receives the profits of the price difference between buyers and sellers and he has thus an incentive to lower the price that is paid to the farmer. If an agent acts as a wholesaler and a commission agent at the same time, this might lead to perverse incentives.[7] First, it might lead to opportunistic behaviour where, in the case the farmer would not be present at the auction, he is charged a commission rate for an auction that never took place. Second, traders would have an incentive to buy at discounts and reduce price information to sellers to a minimum and this in contrast with brokers.

5.3.3.3. Incomplete Information Transmission

Fruits and vegetables are characterised by significant variation in quantity and quality, observable and non-observable. To allow buyers to make right valuation decisions, they need to obtain adequate information on the lot and mechanisms are thus ideally in place to address the asymmetric information problem. As shown above, the majority of the transactions on wholesale markets are done through auctions. Given that often no personal relationships exist between farmers and final retailers and given that there are no repeated transactions between them, which might be an effective way of dealing with information asymmetries (e.g. Kranton, 1996), information transmission is even more important. We discuss what effective practices are on these markets to assess the quality, the quantity, wastage, as well as unobservable characteristics, related to production practices that might influence food safety.

The large majority of retailers believe there are quality differences between the different lots (Table 5.7). 83 per cent of the retailers report to have enough information on quality before placing the bid or doing the transaction. To assess the quality, traders rely exclusively on inspection themselves. 85 per cent of the traders checked the quality and spent an average of four minutes doing so. Quality was checked mostly by looking at and touching the produce. One third of the retailers reported to even have tasted the produce. While only part of the produce could be checked in most transactions, almost all retailers believed that the checked sample was representative.

While modern markets and especially international markets put a high premium on food safety, this is seemingly less the case in these traditional horticulture markets. The use of modern inputs

7. This practice is called "frontrunning" and is prohibited in modern exchange markets (e.g. Chicago Board of Trade).

Table 5.7 **Information Transmission on Quality**

	Unit	Farmers	Retailers
Overall			
There are quality differences between lots			
A lot	%	1	3
A bit	%	93	92
None	%	6	5
It happens that buyer/seller receives/delivers lower quality than paid for			
Regularly	%	0	0
Sometimes	%	25	68
Never	%	74	32
It happens that buyer/seller receives/delivers higher quality than paid for			
Regularly	%	5	3
Sometimes	%	68	52
Never	%	27	45
Last transaction			
Buyer had enough information before transaction	% yes		83
The buyer checked quality himself	% yes		85
If not,...			
... how was quality assured?			
No assurance on quality	%		66
Assurance is based on trust with broker	%		34
If yes, ...			
... time used for checking of quality	minutes		4.4
... way of quality checking			
by looks	%		100
by touch	%		62
by smell	%		7
by taste	%		34
... buyer was able to check whole lot	% yes		34
... if only part of the lot, was it representative?	% yes		90
The buyer knows about production activities (pesticide use, etc.)	% yes		22

is high in horticultural production in India but there is currently little or no transmission in the marketing system for information on the use of inputs (Fafchamps *et al.*, 2007). However, there might be important public health issues related to the lack of proper attention and control of these (Umali-Deininger and Sur, 2007; Athukorala and Jayasuriya, 2003).[8] In our sample, only one fifth of the retailers stated that they were aware of the production activities on pesticide, fertiliser and irrigation water use by the farmers. However, even if they are aware, this does not imply that they do not sell unsafe food as recent research in India shows that there is no premium in traditional retail markets for these un-observable quality characteristics (Fafchamps *et al.*, 2007).

8. For example, Marshall *et al.* (2003) tested fresh vegetables in different production sites and in the main wholesale market in Delhi. They found that 72 per cent of the spinach samples exceeded the Indian Maximum Residue Levels (MRL) and 100 per cent exceeded the Codex MRL level. Kumari *et al.* (2004) found that 26 per cent of their samples of seasonal vegetables contained residues above the MRL levels.

To correctly assess quantities, lots should be weighed. Eighty (80) per cent of the farmers and 73 per cent of the retailers say that they know the exact weight of the lot (Table 5.8). Weighting methods are still old-fashioned as only about one-third of the weighing transactions were done on an electronic scale. When produce is weighed, farmers and traders still complain about "rounding off effects", i.e., where only whole amounts of kgs are paid for. This practice was mentioned by 88 per cent of the farmers and 86 per cent of the retailers. In most cases the advantage of the rounding off was towards the broker or the buyer. Even when no weighing takes place, farmers and retailers might still feel comfortable with traditional units that are in vogue. Only 78 per cent of the retailers reported that they were not satisfied with the assessment of the quantity.

Wastage levels in horticultural marketing are often stated to be high in India (Mattoo *et al.*, 2007) although few good empirical studies are available. Farmers estimated the percentage of the produce that could not be sold at 0.3 per cent of all the transported produce. Retailers evaluated the percentage of wastage at 1.4 per cent of the purchased lot. For the buyers to correctly valuate lots, their expectations on wastage level should be well-informed. If buyers are not able to do this correctly, they might charge uncertainty premiums that are passed through to the farmers. About one-third of the buyers state they do not know the wastage very well of the lot that they will purchase. Nine (9) per cent says that they do know exactly while the majority (57 per cent) knows it approximately.

While a large majority of the traders seem to have enough information to make a reliable assessment on the quality and quantity of the lot, there is however some distrust towards the existing system as illustrated by the asymmetric responses by farmers and retailers towards rewards and payments for quality and quantity. About two thirds of the farmers believe that they sometimes deliver higher quality and quantity than they are paid for while one quarter or less believes that they deliver sometimes lower quality and quantity than paid for (Tables 5.7 and 5.8). The complaints are similar, but in the opposite direction, for retailers.

5.3.3.4. Limited Service Delivery

Two types of transaction costs for retailers as well as farmers can be distinguished, i.e., the monetary costs and the opportunity costs of time to travel to the market as well as to conduct the auction. The average farmer spends almost two hours to travel to the market. The large majority of farmers (94 per cent) use motorised transport to do so (Table 5.9). 38 per cent of the farmers brings produce of other farmers when they come to the market. The transport has usually to be paid for and amounts to almost 10 per cent of the value of the produce that is fetched on the wholesale market. Farmers would spend on average 3.5 hours on the wholesale market. Farmers report that they have to wait for almost 1 hour before the auction can start and that the auction takes on average only 10 minutes. It thus seems that part of their time is used towards other purposes than the sales of their product.

While a significant number of traders and farmers own a phone, it is still not well used for their business. Wholesale horticultural transactions are spot transactions where farmers and retailers show up without much prior contacts (Table 5.9). 95 per cent and 98 per cent of the farmers and retailers respectively reported that they had no contact with the broker before coming to the market. For those that had contact, only a limited number discussed prices with the broker. Most of the price information for retailers and farmers was obtained informally through personal observations or through contacts with fellow farmers or traders.[9]

9. There are currently no functioning commodity exchange markets for horticulture that allow farmers to hedge risks and farmers thus have to live with the uncertainty of price formation during the days of their sales.

Table 5.8 Information Transmission on Quantity

	Unit	Farmers	Retailers
Overall			
There are quantity differences between lots			
A lot	%	1	3
A bit	%	94	90
None	%	5	6
It happens that seller/buyer receives/delivers lower quantity than paid for			
Regularly	%	0	0
Sometimes	%	19	62
Never	%	80	38
It happens seller/buyer receives/delivers higher quantity than paid for			
Regularly	%	5	3
Sometimes	%	64	29
Never	%	31	58
Last transaction			
Buyer has enough information on quantity	% yes		78
Seller/buyer knows exact weight of the lot	%	80	73
If weighted, ...			
..., weighted in front of seller/buyer?	%	80	93
type of scale used is			
... mechanical	%	67	73
... electronical	%	33	27
rounding off of weights	%	88	86
rounding of of weight in seller's advantage	%	16	11
rounding of of weight in buyer's advantage	%	84	89
If not weighted,...			
..., differences between standard units?			
A lot of variation	%	0	8
A bit of variation	%	88	83
No differences	%	12	8
Wastage			
Buyer knows quantity of wastage at purchase			
Exactly	%		9
Approximately	%		57
Not very well	%		34
Does it happen that there is less wastage than expected?			
Regularly	%		46
Sometimes	%		41
Never	%		13
Does it happen that there is more wastage than expected?			
Regularly	%		24
Sometimes	%		28
Never	%		48
Reported level of wastage*	%	0.3	1.4

Note: * wastage level reported by farmer is during transport; wastage level reported by seller is after purchase.

Table 5.9 **Service Delivery**

	Unit	Farmers		Retailers	
		Avg or %	St. Dev.	Avg or %	St. Dev.
Transaction costs					
Auction					
Waiting time before auction	minutes	68.5	96.3	18.8	15.0
Time that auction took	minutes	11.2	9.5	5.6	3.8
Transportation					
Time to travel	minutes	110.5	97.0	23.8	28.4
Time spent on the mandi	hours	3.5	2.2	2.3	1.0
Means of transport:					
motorised transport	%	94		53	
animal	%	5		0	
bicycle	%			4	
on foot/with cart	%			43	
Payment for transport	% yes	91		52	
How much?	Rs per kg	0.4	0.3	0.3	0.2
Payments for transaction					
When paid normally?					
within 3 hours	%	82		78	
between 4 hours and 1 day	%	12		5	
between 2 and 5 days	%	0		2	
after 6 days or more	%	6		15	
Paid in cash		% yes	99		100
Information					
In contact with broker before transaction		% yes	5		2
If yes,...					
..., number of brokers in contact with	number	1.1	0.3	1.4	0.6
means of contact:					
by mobile phone	%	77		100	
by landline	%	15		0	
through representatives in person	%	8		0	
discussed price level with broker	% yes	77		20	
Main source of price information					
Personal observation	%	26		40	
Speaking with regular customers	%			7	
Speaking with other farmers	%	36			
Speaking with other retailers	%			14	
Speaking with CAs/traders	%	24		20	
Screen with price information	%			0	
Observe prices at auctions	%	14		19	

Payments for the transactions are in most cases immediate for farmers as well as retailers. 82 per cent of farmers state that they are paid within three hours after the transaction. The majority of retailers also report to paying immediately for the transaction. However, 17 per cent of the retailers do not pay immediately and are allowed late payments, often until after their produce has been sold (i.e. more than two days). Agricultural trading is largely a cash economy as almost none of the transactions are settled by cheque or other more sophisticated means of payment. Similar results on the importance of spot, unsophisticated and cash transactions have also been found in other poor agricultural economies (Fafchamps, 2004; Fafchamps and Minten, 1999; McMillan, 2002). Finally, while grading is stated to be done on the lot, most of the grading is however done by the farmer and little by the broker, in contrast with brokers in other countries (Gabre-Madhin, 2001).

5.3.3.5. Lack of Competition with the Broker System

Farmers and retailers were asked if they had other potential outlet and procurement sources beside the wholesale market where they conducted their last transaction (Table 5.10). 26 per cent of the farmers report that they could potentially sell their produce at a local village market, 4 per cent to local consolidators and 22 per cent to retailers on retail markets in town. Farmers have thus seemingly little marketing options outside the broker channel. The same holds for retailers. Only 17 per cent could go through a reseller (*masahari*) on the wholesale market or sometimes they got produce delivered at the shop itself by this type of reseller (5 per cent). Few (3 per cent) consider direct sales by farmers to them an option.

While there are few options for farmers and retailers outside the broker/wholesale channel, there are however seemingly a large number of choices within this channel. Farmers state they could potentially choose among 60 brokers that deal in the produce that they are selling. Of these brokers they know five personally. However, they would only use a limited number of them for their transactions, i.e., less than 2 on average (Table 5.10). 57 per cent of the farmers considered only one broker that they would deal with for all their transactions in the last year and a significant number of farmers thus self-select in a specific broker relationship. Often they have a long-term relationship with this broker as they have dealt with the broker from the last transaction on average for almost 10 years. Retailers report a similar large number of brokers to choose from but they usually consider a larger number than farmers that they effectively use for transactions. Retailers placed on average a bid in two auctions and they won on average in one. It thus seems that they only place bids when they are almost sure that they will win it.

The results indicate that there is little choice for farmers and retailers outside the broker system but a lot within. Brokers as a group might thus have an incentive to keep the system as such and try as much as possible to protect their interests from outside competition (Olson, 1965). They seem to effectively do this through the organisation of broker unions that use their powers—and money—to lobby politicians as well as market authorities. This is obvious through the non-respect of the rules of the market authorities. It also shows up in the opposition that traders and brokers were able to successfully organise towards organisations that want to bypass them. For example, traders in the states of Uttarakhand and Uttar Pradesh succeeded in delaying the investments by modern retail in food markets by calling for large strikes. It is especially interesting that the modern retailer Reliance, that had a clear strategy of wanting to bypass brokers and to buy directly from farmers, got most of the heat of these protesters.[10] On the other hand, large schemes such as the e-choupal model, that used brokers while bypassing the regular wholesale markets, were easily accepted (Goyal, 2008).

We find that the margins between the seller and buyer are high (between 13 per cent and 26

10. Interestingly, farmers also staged protests as to keep modern retail procurement there. They however lost.

Table 5.10 Choices in Outlets and in Brokers

	Unit	Farmers		Retailers	
		Avg or %	St. Dev.	Avg or %	St. Dev.
Outlet options					
Consolidator	%	4			
Village market	%	26			
Broker	%	100		89	
Wholesaler at wholesale market	%			17	
Wholesaler delivering at shop	%			5	
Direct marketing at *mandi*	%	22		3	
Visit to other market	%	16		2	
Choice between brokers					
Perceived number of brokers that sell this product	number	62.3	42.5	46.7	21.0
Number of brokers you know that sell this product	number	5.3	4.8	6.1	5.0
Number of brokers that you use	number	2.0	1.7	3.8	2.9
Number of brokers considered today	number	1.2	0.5	1.4	1.1
Number of auctions attended	number			4.0	3.2
Number of auctions where this retailer placed a bid	number			2.1	1.3
Number of auctions where this retailer won the bid	number			1.2	0.6
Frequencies of transactions					
Number of transactions through brokers					
this season	number	9.6	9.7		
last year	number	26.1	27.2		
last two weeks	number			9.0	2.7
Number of brokers used for these transactions					
this season	number	1.6	0.9		
last year	number	1.8	1.2		
last two weeks	number			3.5	2.6
Number of brokers used last season (farmers)/last 2 weeks (retailers)					
One	%	57		17	
Two	%	33		19	
more than two	%	10		64	
Time dealt with the broker of last transaction	years	9.7	9.1	9.2	7.1

per cent), regulations are not respected, the wholesale trader and broker role are being confused, information transmission is incomplete, transaction times are long and outside competition with the broker system is limited. This leads seemingly to lower sales prices for the producer and higher purchase prices for the retailer. However, farmers and retailers do come to these markets and the successful sustainable development of alternative marketing institutions has been limited. In the New Institutional Economics literature, different authors (e.g. North, 1990; Fafchamps, 2004) have shown that institutions are not passive entities and do adjust to the constraints of particular economic environments. In the next section, we will look in more detail at the services that brokers provide and how the broker institution has been able to persist in agricultural markets.

5.3.4. Reasons for Resilience of the Broker System

5.3.4.1. Low Transaction Costs

Physical wholesale markets have been prevalent throughout history as the most widespread way to bring buyers and sellers together (McMillan, 2003).[11] To get at their preferences for these wholesale markets, farmers and retailers were asked why they chose the broker/wholesale channel above other potential outlets (village markets or direct selling to retailers). The results are presented in Table 5.11. Farmers prefer using the broker channel because of the good prices that are offered, the large quantities that can be disposed of and the

Table 5.11 Search Costs

	Unit	Farmers	Retailers
Advantages of brokers considered to be 'very important'			
Overall			
"Prices are better"	%	63	79
"I can sell/buy in large volumes"	%	70	46
"Access to input advances and credit"	%	26	
"I can defer payments"	%		15
"I have access to loans"	%		4
"The transaction is fast"	%	60	62
"Payments are immediate"	%	67	
Last transaction			
"He finds lots of potential buyers/sellers"	%	46	34
"He offers better prices"	%	55	64
"He offers higher quality"	%		66
"He gives seasonal input advances"	%	26	
"He allows me to defer payment"	%		16
"He offers loans in case of need"	%	21	4
"I have the habit"	%	50	29
"He has quick transactions"	%	63	58
Regularity of supply/outlet			
It happens that buyer/seller is unable to sell/buy this product			
Never	%	89	54
Occasionally	%		46
Often	%	0	0
If it happens, buyer/seller...			
... takes it home/sells next day	%	91	
leaves it at wholesale market	%	4	
sells to a wholesaler	%	4	
trades another product	%		75
does no business today	%		7
performs wage labor today	%		5

11. Depending on the nature and the size of transaction costs, some people might however forgo participating in these markets (Fafchamps and Hill, 2005).

relatively quick transaction. While some farmers prefer the broker channel because it gives them access to credit and input advances, its number is limited (25 per cent of the answers). Similar trends show up in the answers of the retailers. Late payments and access to insurance is only mentioned by a limited number of retailers as the main reason why they rely on brokers.

As retailers, and especially farmers, seemingly select long-term relationships with specific brokers, they were asked why they chose the specific broker in their last transaction. Again, same trends show up. While habit formation is part of the explanation of going through a specific broker (50 per cent and 29 per cent of the farmers and retailers respectively state this to be very important), most of the decision is however based on a reduction of the search costs and on obtaining the best price possible. Interlinked market services provided by brokers are especially less important for retailers. Only 15 per cent and 4 per cent of the retailers would choose a particular broker because he thinks it very important that they provide the option of late payments or access to credit respectively. However, these numbers are as high as 26 per cent for the provision of input advances and credit in the case of farmers. To this, we turn next.

5.3.4.2. Access to Insurance and Credit

A common problem for rural agricultural economies is the prevalence of different types of shocks for which especially poorer households might be ill-prepared. Given the lack of formal insurance mechanisms, households must often rely on social capital and selling of assets to deal with these shocks and only those households that have these informal insurance mechanisms are able to successfully smoothen their consumption (e.g. Dercon, 2005; Rosenzweig and Wolpin, 1993; Townsend, 1994).

Some farmers—and a limited number of retailers—use the broker for insurance purposes. 39 per cent of the farmers report that in case of need, the broker would give loans to the farmer for sure (Table 5.12). 18 per cent of the farmers thus think that he would "probably" do so. While more than half of the farmers think that they could rely on the broker in case of need, only 22 per cent of the farmers have ever received a loan from the broker with whom they dealt in the last transaction. 20 per cent of the farmers received a loan in the last 5 years. The average value of the loan was Rs. 8263 (more than US$200) or about twice to four times the value of the last transaction. However, farmers seldom rely exclusively on brokers for access to credit as 96 per cent of the farmers report to have alternative sources of credit. These include formal banks (46 per cent of the farmers) but more importantly friends and family (78 per cent). Access to credit through brokers is less important for retailers. Only 7 per cent of the retailers believe that the broker would give loans for sure in case of need and only 2 per cent of the retailers report to ever have received a loan from a broker. Retailers also can rely on other options for access to credit as stated by 95 per cent of the retailers.

5.3.4.3. Access to Input Advances

A second advantage for farmers of working with specific brokers might be the access to input advances. This access proves problematic for some—and often the poorest—farmers all over the developing world, often due to seasonal liquidity constraints (Dercon and Christiaensen, 2007). 22 per cent of the farmers received this year an input advance of the broker that they dealt with in the last transaction. For half of the farmers, this advance was in kind, more specifically seeds. No fertiliser or pesticides were given in kind to any farmer in our sample. The value of the seed advances was evaluated at Rs. 6,115. The quality of the seeds was evaluated by most farmers to be good (68 per cent) and almost half of the farmers think that they could not find the same quality

Table 5.12 **Credit and Insurance Services Provided by the Broker**

	Unit	Farmers		Retailers	
		Avg or %	St. D.	Avg or %	St. D.
Overall					
Number of brokers that farmer/retailer could obtain a loan from	Number	1.1	1.1	0.6	1.2
Broker used in last transaction					
The broker gives loans in case of need					
yes, for sure	%	39		7	
probably	%	18		22	
no	%	44		71	
Buyer/seller ever received a loan from this broker	%	22		2	
Number of loans received in the last five years	Number	1.9	1.1	2.4	1.7
The value of the loan - mean	Rs	8263	9429	21500	38464
Buyer/seller has other sources of loans	% yes	96		95	
If yes, from...					
...bank	%	46		17	
...friends/family	%	78		81	
...others	%	2		14	

themselves at the same price. On the other hand, they do not believe that productivity of their vegetables was higher because of the use of these inputs. Almost none of the brokers (11 per cent) monitored the use of these seeds. Only 4 per cent of the farmers reported to have to pay interest on this advance in kind.

The other half of the farmers that received input advances, received inputs in cash. The amount was on average almost Rs. 7,000. Only one farmer that received this advance in cash was asked to pay interest on it (Table 5.13). Surprisingly, only 6 per cent of the farmers state they would receive every year input advances from this broker. It thus seems that they would have to make the case every year for the need of this money. We also asked farmers what the broker would do if the farmer would not pay back the input advance. As is usual the case in this type of markets, formal enforcement mechanisms are not well relied upon (e.g. Fafchamps and Minten, 2001; Bigsten *et al.*, 2000; McMillan and Woodruff, 1999). Farmers report that it is very unlikely that the broker will go to the market authorities, to the police or to the court. In case of default, he will however not work anymore with this farmer, other brokers will hear about it and some brokers might use peer pressure in the village.

5.3.5. Broker Economics

In an efficient economy, brokers and traders would be rewarded for their costs and any risks they may take. Gabre-Madhin (2001) distinguishes five roles for the agricultural brokers in the case of grain trade in Ethiopia: (1) provide price information; (2) arranging logistics of delivery; (3) grading; (4) determine the market price; (5) and matching buyers and sellers. Based on our case study, the role of brokers here is seemingly more limited and different as they do not provide price information; are not involved in delivery logistics which is the responsibility of the farmers; and they do not do—or do only limited—grading. Brokers seemingly perform two distinctive activities

Table 5.13 Input Advances Provided by the Broker

	Unit	Farmers	
		Avg or %	St. Dev.
Overall			
Number of brokers that farmer received input advances from this season			
From none		%	78
From one broker		%	22
From two brokers		%	1
Broker used in last transaction			
Farmers that received an input advance		%	21
Of those that received input advances, ...			
...input advances were partly in kind		% yes	55
Details on input advances in kind			
Farmers that received seeds		%	96
Value of seeds received - mean	Rs	6115	7769
"Quality of seeds"	% good	68	
"Could have access to market access for same quality"	% yes	56	
"Productivity of these seeds is higher than if bought myself"	% yes	22	
"Farmer can find inputs himself at a cheaper price"	% yes	15	
"Broker checks if inputs were used on farmer's fields"	% yes	11	
"Interest payments on this advance"	% yes	33	
...input advances were partly in cash	% yes	55	
Details on input advances in cash			
Amount of cash received - mean	Rs	6982	9546
"Interest payments on this advance"	% yes	4	
...the farmer receives input advances every year	% yes	29	
...what would happen if these input advances were not paid back?			
"Broker will not work with me anymore"	% yes	84	
"Broker would complain to the market authorities"	% yes	20	
"Broker would complain to the other brokers"	% yes	77	
"Broker would use social pressure in the village"	% yes	55	
"Broker would bring me to the police or court"	% yes	2	

in the case in the case of Uttarakhand: they organise auctions where they bring sellers and buyers together and they give input advances and access to credit to a number of farmers. Below, we elaborate on the costs and rewards for these activities.

One justification for the broker margins could be rewards to search costs (as shown by Gabre-Madhin, 2001 in the case of brokers in Ethiopian grain markets). However, the costs of bringing sellers and traders together is limited in this type of horticultural trade given that buyers and sellers of these commodities, that are in the majority locally traded, are all coming to the same wholesale market, as they are actually required by law to do so.

Table 5.14 **Determinants of Loans and Input Advances Received by Farmers (Heckman Model)**

		Last loan received		Input advances received	
		Coefficient	z-value	Coefficient	z-value
Determinants of log(Amount in Rs)					
household characteristics					
member of scheduled or backward caste/tribe	yes=1	-0.438	-1.03	-0.181	-0.49
years of education	log(number)	-0.024	-0.08	-0.238	-1.40
household size	log(number)	0.019	0.43	-0.062	-1.34
holder of Below Poverty Line (BPL) card	yes=1	0.333	1.28	0.252	0.86
production characteristics					
area cultivated of the studied crop	log(area)	0.501	3.69	0.435	2.98
intercept		7.874	7.59	8.684	8.35
Selection equation					
household characteristics					
member of scheduled or backward caste/tribe	yes=1	-0.245	-0.93	-0.029	-0.12
years of education	log(number)	0.498	3.01	0.061	0.47
household size	log(number)	0.049	1.30	0.030	0.86
holder of Below Poverty Line (BPL) card	yes=1	0.428	1.67	0.391	1.61
production characteristics					
area cultivated of the studied crop	log(area)	-0.080	-0.62	0.082	0.65
market characteristics					
market of dehradun	yes=1	-0.301	-1.12	0.013	0.05
number of visits per year	log(number)	0.388	2.39	-0.036	-0.24
distance to the market	log(hours+1)	-0.746	-3.09	-0.814	-3.33
intercept		-1.449	-2.01	0.218	0.34
Number of observations		237		237	
Censored observations		185		192	
Uncensored observations		52		45	
Wald chi2(5)		16.46		15.43	
Prob>chi2		0.006		0.0087	

Access to loan and input advances and the amount given are presumably rational economic decisions by the broker (Bell and Srinivasan, 1989; Crow and Murshid, 2003). To explore what the determinants are that drive access to credit from brokers to farmers, we rely on data of actual loans given. We use a Heckman strategy where we estimate in a first stage the likelihood that a farmer was a beneficiary of loan in the last five years and then estimate in a second stage how much he received. The distance from the trader to the wholesale market serves as an instrument in the selection equation.[12] The results are shown in Table 5.14. Access to loans is seemingly not well linked with need as poorer farmers have equal access to loans as richer ones (as measured to land cultivated or access to a BPL card). This confirms earlier results of Bell and Srinivasan,

12. Distance to the wholesale market should be a determinant of the costs of recuperating the money in case of default but conditional on receiving a loan, it should not affect the amount of the loan.

1989 where they found credit-marketing linkages in India to be strong in the state where larger farmers dominate.[13] More highly educated farmers are more likely to receive a loan. Farmers that live further from the market are significantly less likely to receive a loan while farmers that visit the market more often are more likely to receive a loan. This is seemingly due to differences in the monitoring costs. Conditional on receiving a loan, farmers who cultivate more land of the vegetable under study, receive significantly bigger loans. A doubling of the area increases the amount by 50 per cent.

We follow the same Heckman strategy in analysing the determinants of the use of inputs advances (we aggregated advances in cash and in kind). As for the case of access to credit, distance between the wholesale market and access to input advances show a significant negative relationship in the selection equation (Table 5.14). The further the farmer is located, the less likely he is to receive input advances. Conditional on receiving input advances, the received amount is again significantly related with the land that the farmer cultivates of this crop. A doubling of the area increases the amount of input advances by 43 per cent. These results thus indicate that brokers provide this access to interlinked markets rationally, not based on household characteristics of the farmer but on monitoring costs and the ability to recuperate their advances.

The costs for allowing access to credit and input advances to farmers are twofold. First, there are the interest costs that the broker/trader does not charge to the farmer and second, there are the risks of default. Using the numbers coming out of the previous sections, we simulate what the returns are on these risky investments and calculate the internal rates of return for different default rates and for different broker rates (Figure 5.3). The results show that even if default rates

Figure 5.3 **Returns to Investments for Credit and Input Advances**

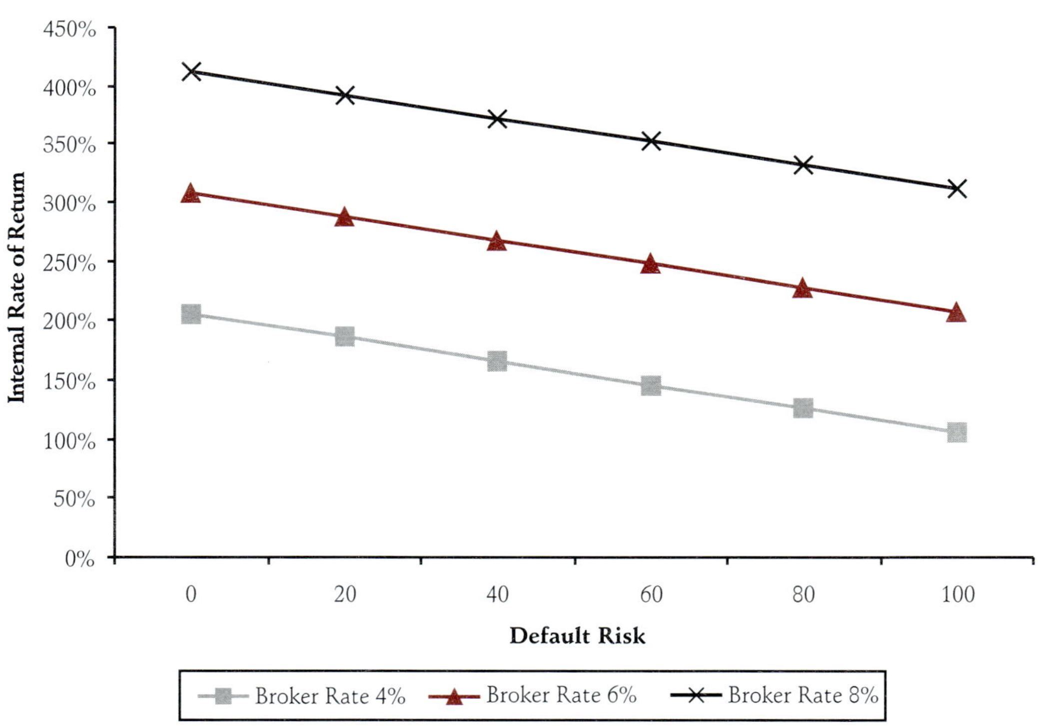

13. Crow and Murshid (2003) find in Bangladesh that social power is an important determinant of access to loan and that poorer household get loan but on less advantageous conditions.

would be a 100 per cent—implying that none of the farmers would pay back any of the input advances or loans given to them—internal rates of returns on these investments would still be above 100 per cent for a low broker rate of 4 per cent, significantly higher than interest rates paid in informal or formal credit markets. More realistic broker rates and low default risks push rates to between 200 per cent and 400 per cent.

These results suggest that the major role of the brokers would not be to pay for these interlinked market services as with these types of returns, other providers would presumably emerge that would deliver these services. Part of the returns is thus linked to other services that he provides. It is also likely that part of the broker rates might be rent rewards due to licenses and space limitations in the wholesale markets. The efficiency of the whole marketing system of horticultural crops might then be improved by encouraging extra competition with the broker system and/or a more efficient set-up of these auctions potentially not run by brokers but, for example, by appointed auctioneers as is done in some grain markets in India (Banerji and Meenakski, 2004).[14]

5.3.6. Conclusions and Implications

In this case study of traditional wholesale markets in Uttarakhand, we find that marketing regulations are ineffective, that there is confusion on the function of brokers and wholesalers, that information transmission on quantity and quality between buyers and sellers is incomplete and that the monetary costs for transactions on the wholesale market amount to at least 12 per cent of the wholesale price. Despite these problems, the broker system has been prevalent in Indian agricultural markets due to its ability to interlink credit and insurance markets for farmers to the agricultural output market. However, this latter function seemingly only explains part of broker margin.

The interviewed farmers and retailers state they are still largely satisfied with the current system, seemingly because the wholesale market reduces search costs and allows for access to a market place with lots of sellers and buyers. However, the set-up of the broker system could likely be improved. Brokers in any market are rewarded for reducing the search costs of sellers and buyers (see e.g. Hsieh and Moretti, 2003; Gabre-Madhin, 2001). It is not evident to what extent agricultural brokers in India deliver these services as buyers and sellers have few marketing options outside the wholesale markets and both will show up physically there (as required by law). As services for the reduction of search are thus limited, the broker system as currently practiced might then just be an expensive way to conduct auctions. Alternative cheaper options could be explored to conduct such auctions.

The broker system further seems in its current form not well adapted to deal with the upcoming food safety and traceability demands in modern local or export horticultural markets. Buyers obtain little information on unobservable characteristics of the product and the regulated broker system—through its anonymity—is seemingly a hindrance towards the development of closer vertical coordination and better information transmission between producers and buyers which is much needed in modern markets. Interestingly, this movement is the reverse of what the initial development of markets entailed where the move away from personalised to anonymous transactions was a condition for success, specialisation and economies of scale (McMillan, 2002; Fafchamps and Minten, 1999; Fafchamps, 2004).

---

14. Other authors have come to similar conclusions recently. For example, Goyal (2008) found that producer prices of soybeans in Central India increased by 1 per cent to 5 per cent on regular wholesale markets after the introduction of alternative marketing channels where soybeans could be procured directly from farmers. She explains this upward pressure by access to increased information by farmers and the pressure that this put on traditional traders.

6 High-value Crop Chain Analysis

6.1. Introduction

The commercialisation of agriculture (in input as well as output markets) has been the cornerstone of economic development for many developing countries (von Braun and Kennedy, 1994). A shift away from subsistence agriculture leads generally to better welfare and nutrition. Our case study in Uttarakhasi shows that welfare conditions improved significantly since the production of off-season vegetables for the market (Box 6.1). Other authors have found similar

> **Box 6.1 The Success Story of the Cultivation of Off-seasonal Vegetables in Uttarakhand**
>
> "Before, we ate only meat in the cold months; now, we eat it during the whole year. Before, milk was only consumed if people had buffaloes; now, everybody has at least one Jersey cow. Before, wheat and rice were luxury goods and only given to guests; now we buy it from the market. Before, it used to be shameful to buy food from the market; now, it is a source of pride" Focus group interview farmers 22/07/07
>
> The steep reaches of the Rawain valley of Uttarakhand are an unusual site for the humdrum of trucks and tempos, racing ferociously to connect the scatter of villages in the hills with Dehradun, Delhi, Meerut, Kanpur and Lucknow. Over the last five years or so, the area has emerged as a booming sourcing centre for off-seasonal vegetables, meeting increasingly rising demands for vegetables between June and August in these centres and beyond. 35 trucks per day, laden with vegetable produce destined for retail/wholesale markets in the plains, chief among these being tomatoes, French beans, aubergines, capsicum, green peas, cabbages and cauliflower. The valley currently produces an estimated 10,000 tonnes of off-seasonal vegetables between the months of June and August. What began as a participative venture between the Himalayan Action Research Centre (Himalayan Action Research Centre (HARC), a local NGO working towards sustainable livelihoods in remote hill communities), and Mother Dairy, (a government-supported, highly successful organised food retail establishment catering to fruit and vegetable demand in Delhi) has transformed cultivation and livelihoods patterns for the valley as a whole.
>
> Our interactions and investigations reveal a dramatic rise in the cultivation of off-seasonal vegetables from about 5 per cent of cultivable area to about 50 per cent for Mother Dairy-HARC partner families as well as others, all over a period of 8–10 years. Ten years ago, 5 per cent of the farmers in the area would grow off-seasonal vegetables but now almost all farmers do so. The transformation from subsistence to a market economy is evident in the increased cultivation of vegetables on a commercial basis (as against previously for self consumption alone) as well as the increasingly reported consumption of products such as milk, fruit and processed grains. Before, farmers used to grow, store and eat their own food. Now, they increasingly sell cash crops at a high price and buy staples at a low price. All this has been possible because of the market-driven stimulus in the area, initiated by Mother Dairy, but increasingly propelled by scores of private procurers.

Table 6.1 Value Chain Issues by Crop

	Product				
	Temperate fruits	Potato	Off-season vegetables	Organic crops	Medicinal and herbal crops
Production					
Number of producers involved	Medium	Large	Medium	Small	Small
Input use					
- irrigation requirements	Little	Little	Large	Depends	Depends
- modern fertiliser use	Little	Medium	Large	None	None
- pesticide use	Medium	Medium	Large	None	None
- capital requirements	Large	Little	Little	Medium	Little
- labour intensity	Medium	Large	Large	Large	Little
Production risks					
- wildlife	Large	Low	Medium	Medium	Low
- climatic risks	Large	Low	Low	Large	Low
- plant disease risks	Low	Medium	Medium	Large	Low
Marketing					
Availability market outlets	Large	Large	Large	Few	Few
Price rewards	Variety/Size	Freshness	Shelf-life	Type Certification	Type of Crop
Wastage	Medium	Low	Large	Large	Low
Packaging material	Wooden Boxes	Bags	Plastic Crates	Depends	Depends
Commission rates to farmer	8%-10%	4%-5%	6%	Not on mandi	1%
Storage possibilities of crop	Yes	Yes	No	Depends	Yes
Farmers' share in final price	Medium	Large	Medium	Large	Small
Demand					
Price risks					
- interannual	Small	Large	Small	Depends	Depends
- seasonal	Medium	Large	Small	Depends	Depends
Local market size	Medium	Large	Large	Small	Small
Local market size trend	Stable	Growing	Growing	Growing	Stable
Trade	Importer	Exporter	Non-traded	Exporter	Exporter

results in Uttarakhand.[1] Especially the commercial high-value agricultural sector in Uttarakhand has significant potential and the stimulation of its growth is high on the government's agenda.

However, there are still significant constraints in the structure of current commodity chains that do not allow farmers to fully exploit the opportunities offered in the changing market environment. Currently, there are only pockets of change and more could be done to improve the growth of high-value agriculture. Table 6.1 gives an overview of the structure and some issues in the different value chains. In the annex of this report a detailed value chain analysis is done for important high-value crops that are currently grown in the state. The study of these value chains

1. For example, Bhadani (1998) found in the Nainital district that a shift from traditional agriculture to market-oriented vegetable farming improved the income of farmers by 172 per cent and the employment opportunities by 169 per cent.

Table 6.2 Fruits Suitable for Different Altitudinal Zones

Zone	Valleys	Suitable for north, east and west facing slopes	Suitable for south facing slopes
2400 m and above	Apple, pear, plum and cherry	Apple, pear, plum and cherry	Apple, pear, plum and cherry
1800–2400 m	Apple, pear, plum and apricot	Apple, pear, plum and cherry	Peach, plum, apricot, pecannut, chestnut, walnut and low chilling varieties of apple
1200–1800 m	Peach, plum, apricot, pecannut, low chilling varieties of apple and pears	Peach, plum, apricot, pecannut, chestnut, walnut, apple and pear	Citrus and peach
Below 1200 m	Mango, citrus, guava, strawberry and peach	Mango, citrus, guava, peach, papaya, banana	Mango, citrus, guava, papaya, banana and aonla

Source: Gupta *et al.*, mentioned in Rawat and Mishra (2001).

is not an indication that these particular crops should be promoted in particular. The idea was to choose important crops of the high-value fruits and the vegetable groups and document what the issues are which appear in the structure of these value chains. The three products chosen were potatoes, tomatoes and apples. Using quantities produced as the yardstick, potatoes and tomatoes are the most important vegetables in the state and apples are the second most important fruit in the state after mango (but production of apple is more geographically spread than mango, which is the most important crop). Given the important perceived potential and the predominance in policy discussions, organic crops and herbal plants were also selected as additional value chains to be looked at in more depth. A write-up of the structure and issues of their value chains is given in Annex I through V.

We now give a brief overview of the different high-value crop chains in Uttarakhand as well as a summary of needed investments as they emerge from the interviews with stakeholders as well as from the reading of the literature. The different sectors that will be discussed are horticulture, organic crops, medicinal, aromatic and culinary herbs, floriculture and seeds. We will finish the chapter with the results of simulations on the impact of interventions in the value chain.

6.2. An Overview of High-value Chains

6.2.1. Horticulture[2]

The varied climate of Uttarakhand makes it an ideal location for growing temperate, sub-tropical and tropical fruits and off-seasonal vegetables that fetch a high price in the domestic urban markets and international markets. For example, Table 6.2 shows the types of fruits that can be grown in the different zones of the state, distinguished by altitude. As indicated earlier, the consumption pattern of the Indians is quickly changing and the demand for fruits and vegetables is increasing over time which makes this sector an ideal sector to promote, with likely significant impacts on poverty alleviation, especially so for the labour-intensive vegetables.

The climate in Uttarakhand allows the hilly regions to supply fruits and vegetables during a period when few other competitors are on the market. Prices for off-season vegetables are significantly

2. For a detailed overview of the value chains in horticulture, see Annex I (apples), Annex II (potatoes) and Annex III (tomatoes).

Figure 6.1 **Seasonal Patterns in Potato Arrivals in Hills and Plains in Uttarakhand (April 2006-March 2007)**

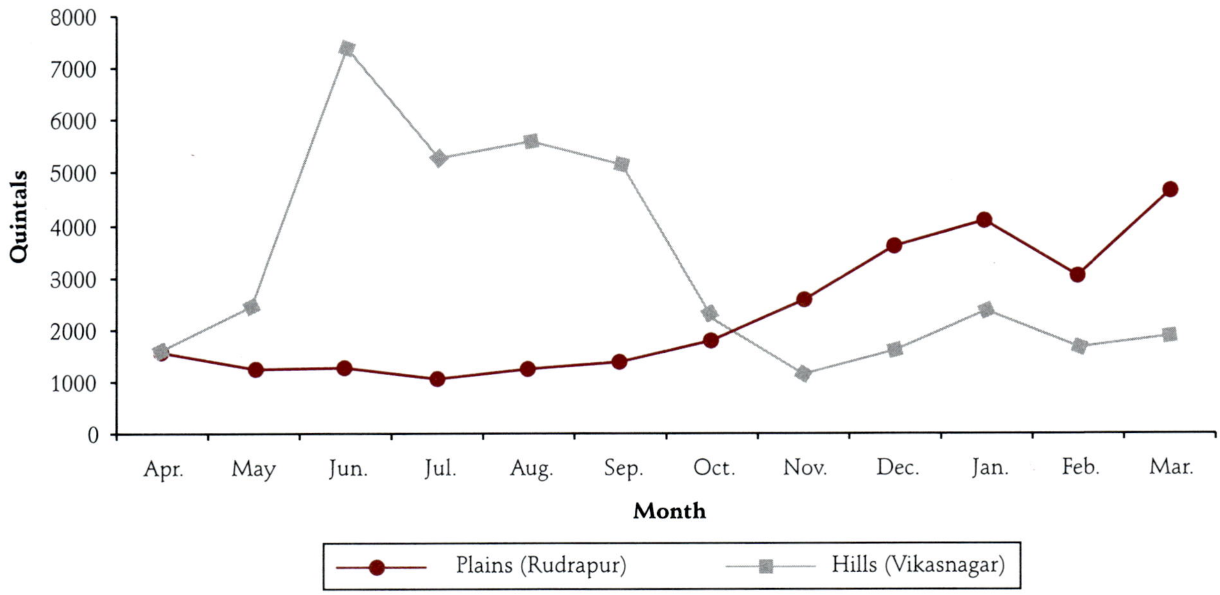

Source: Calculations based on Agmarknet data.

Figure 6.2 **Wholesale Prices on Markets in Uttarakhand and Delhi (April 2006-March 2007)**

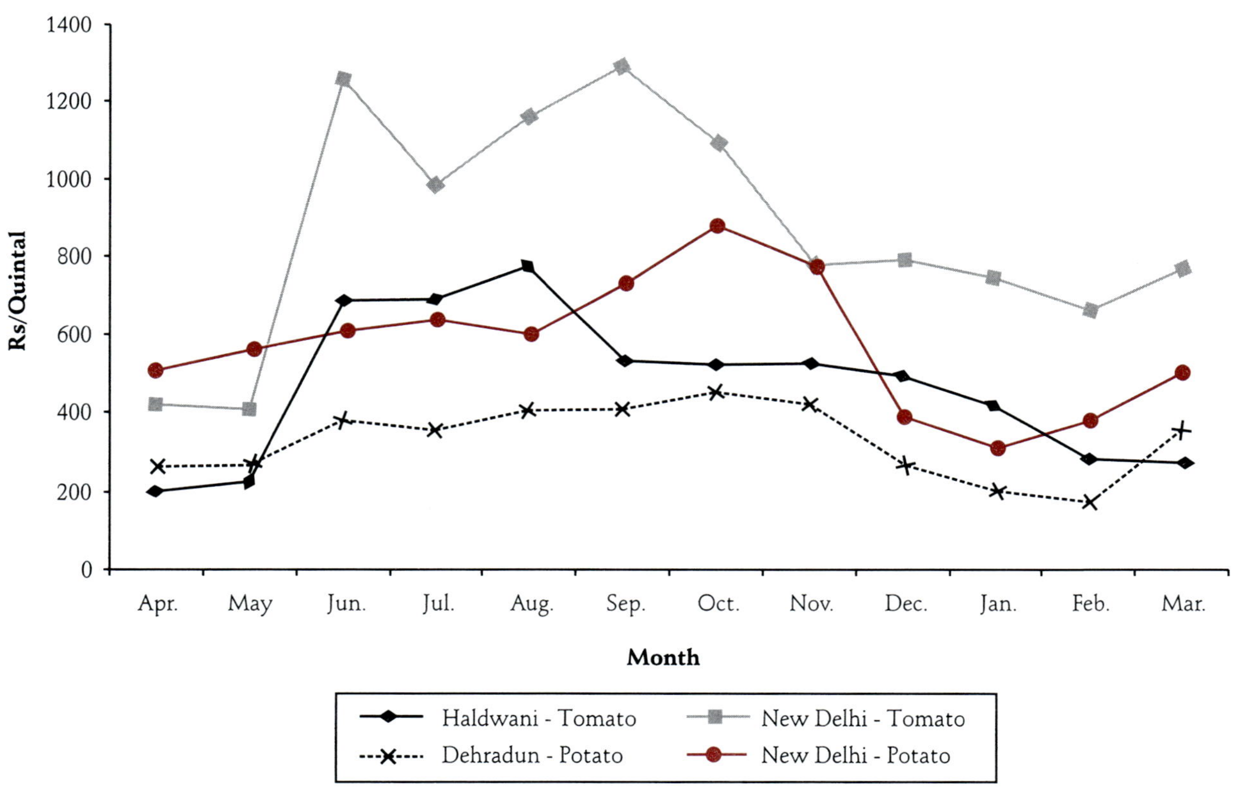

Source: Own calculations based on Agmarknet.

higher than during the main season (Figure 6.2). Even if storable products are available on agricultural markets during the production periods in the hills, the products from the state of Uttarakhand might still have an advantage and obtain a price premium due to its freshness (for example at the time of our survey, hill potatoes would fetch a premium of 55 per cent over cold storage potatoes in the New Delhi market). Fruits can also be supplied at a different time than products in other hilly states, often reaping a price premium for that advantage (as for example in the case of apples).

When we look at the demand side of horticultural products, we find significant variation in price and consumption for different products over time. We find large variation in prices as well as consumption (Chapter 3). Some vegetables (such as potatoes) and fruits (apples, papaya and guava) have shown significant price increases while others saw significant declines (especially onions). Per capita consumption has been on the increase for most of the fruits and vegetables over the last fifteen years but it has been more pronounced for products such as potatoes (+25%), tomatoes (+42%), cabbage (+107%) and bananas (+71%).

To capture the market, different constraints exist, in production as well as marketing in the state of Uttarakhand. Production constraints for horticulture production are multiple and relate to old and senile orchards, lack of quality planting material, poor and faulty nutrition, lack of pollinisers and pollinators, vagaries of weather, incidence of pests and diseases, faulty inter-cultural practices, low planting densities, unavailability of horticultural inputs, lack of technical know-how and poor linkages between research and extension institutions. A main problem that especially limits the development of vegetable production in the hills, is the lack of irrigational infrastructure (Kar, 2006). Although Uttarakhand gets very heavy rains during the monsoon season, the topology of the area and the nature of the soil do not permit the soil to retain this moisture. Parts of the state are also drought-prone with inadequate rainfall.

While the farmers would benefit enormously in the long-run by planting other varieties that fetch higher prices, better post-harvest technologies and value chain management would also allow farmers to obtain a better price in the short-medium term. While it is difficult to pinpoint specific fruits and vegetables to invest in, some stakeholders however indicated that some crops were more suited than others. Pears and walnut were two fruits crops that were estimated by different stakeholders to have good potential in the hilly areas of Uttarakhand as well as to have significant opportunities in the market. Investments towards a more efficient value chain would help in the promotion of these crops.

6.2.2. Organic Crops[3]

It is a major objective of the government to stimulate organic production in the state. One major strategy is to increase the number of farmers under organic certification. The state target for 2010 is the certification of more than 100,000 farmers. Farmers are mostly eager to participate in such participation schemes as it allows them to potentially capture part of the high organic price premium (IFAD, 2005). To evaluate the incentives for organic production in the state, it is important to try to understand the existing price premium in this organic market. There seem to be two major reasons, the first linked to higher production costs and the second due to niche market issues.

First, organic production might be more costly than traditional production methods for several reasons: 1/ Most authors state that organic practices lead to decreases in productivity, especially

3. For a detailed overview of the value chain, see Annex IV.

during a transition period of two to three years. Thereafter, they recover. Sometimes it is not the productivity of the crop itself that declines but the crop that comes after it (as reported in the case of rice and wheat) though these costs may be minimal in those parts of the state that typically do not use much chemical fertilisers and pesticides;[4] 2/ Wastage might be higher during storage when no chemicals can be used (as stated by different organic processors); 3/Organic certification is an expensive and time-consuming process that has to be paid for. Different certification agencies are active and the costs associated with certification seem related to the type of markets one is aiming for. Local, cheaper certification as these provided by the Uttaranchal Organic Certification Agency (UOCA) seems to be sufficient to get access to the local organic market while certification by international agencies such as Skal International (based in Mumbai) or Onecert Asia (based in Jaipur) are required for export markets.[5]

Second, prices might be higher because this market is a niche market where sellers and buyers have not yet found each other well. For example, our field research showed that from 80 certified villages in the Almora district, only around 20 per cent of all fruits were sold to the organic market segment while 80 per cent was sold in the regular *mandi* without any premium. On the other hand, some organic basmati rice producers would receive a premium that is 100 per cent higher than the price of regular basmati rice, seemingly significantly higher than the adjustment and certification costs. Prices for organic produce might then be relatively high for some producers due to the high willingness-to-pay by a limited number of consumers. Once this market has grown (which it is expected to do), there is no reason, as basic economics would predict, that the price premium that is offered to producers should not reflect the extra production and certification costs that organic farmers face in well-established markets (IFAD, 2005). Large market premia for organic produce that do not reflect production costs are thus seemingly not sustainable in the long-run. However, premia above production and adjustment costs might exist in the medium-run because of the rapidly expanding organic food demand (as stated by multiple stakeholders in this area) as supply might have difficulties catching up (given the lag in certification).

This does not mean that organic agriculture does not have a role in the long-run, but organic production should be justified by good economic incentives in the long-run. Organic production is disadvantaged due to the subsidies given to modern inputs (at a huge cost for the government). It makes little policy sense to aim for organic production processes and then subsidise modern technologies such as chemical fertiliser and pesticides. This creates biased economic incentives towards the adoption of these modern technologies. Despite this bias, some compelling reasons do exist for the promotion of organic production in Uttarakhand as investments in organic production practices might especially have large advantages in hill production environments characterised by soil erosion. In such an environment, organic production might contribute enormously to soil health.

It seems that sensible development interventions for organic production must focus on providing support in the form of extension services. The Organic Commodity Board for example is understaffed, and would need further assistance to scale up its operations and train farmers on organic practices. In a recent detailed case study of organic basmati rice farmers in Uttarakhand,

---

4. For example, in the case of basmati rice, rice productivity seems little affected (but it is the productivity of the wheat, the crop grown after rice on the same field that might be significantly lower in productivity). In the case of tomatoes, farmer focus groups reported a drop in yields between 25 per cent and 30 per cent. It seems that especially fungus diseases are a major issue in organic vegetable production and there are currently no effective organic fungicides available.

5. Costs borne by the demander often include transport costs and accommodation of the inspectors of the certification agency and yearly inspection costs per acre that differ for the market that the demander is aiming for (for example, Japanese, European or US markets).

Alam (2007) finds that most farmers complain about access to information and know-how. More attention to research on developing better and more varieties of organic seeds, pesticides and fertilisers should be a thrust area. With respect to crops to focus on, talks with large traders that sell organic products in retail markets in Delhi revealed that there is currently an important demand for organic spices (for example, chillies is an upcoming market) and pulses that they find hard to fulfil. Focusing on these crops might, in the short-run at least, yield assured buyers.

6.2.3. Medicinal, Aromatic and Culinary Herbs[6]

Only a limited number of farmers are involved in the cultivation of herbs and medicinal plants in the state of Uttarakhand. However, they are seemingly relatively more important in the most remote and high-altitude areas of the state, implying that they might be important crops to focus on for poverty alleviation given the high incidence of poverty in this area. Given the largely informal nature of this trade, it is unfortunately difficult to obtain exact statistics on their importance in the state of Uttarakhand.

Views on the potential of this sector differ greatly. Farmers interested in the cultivation of herbs and medicinal plants might receive relatively few rewards for it and they have thus few incentives to convert. First, the competition between collection and cultivation depresses the cultivation price. Second, the illegal nature of the majority of the trade pushes prices at the farm level even further down. Third, these low prices of the herbs and medicinal plants at the farm level give the farmer little incentive to grow herbs compared to, for example, off-seasonal vegetables. For example, it is estimated that one would need three times the wholesale price of medicinal plants in Delhi to make growing medicinal plants profitable compared to a combination of potatoes and peas. This does not even take into consideration the uncertainty of crop production and the price uncertainty as production often only starts three years after planting (Alam, personal statement).

However, it is hard to make general statements on the sector. The particular benefits of large scale trade of certain crops may not be as high as others and there needs to be a case by case evaluation of what may be viable. While many medicinal plants, with long gestation periods and that grow in the wild, cannot be viably cultivated, the very visible success of culinary herbs exports indicates that there might be lucrative niche markets that some enterprise can effectively tap into. The challenge lies in identifying those specific niches where strong market linkages may be forged. With respect to crops to focus on, talks with informed stakeholders indicated that they felt that lavender and stevia are crops with large upcoming demand.

Development interventions in this sector could take on many roles. The establishment of collection centres, on perhaps a smaller scale than the existing herbal *mandis* but reaching the remote areas could be a welcome step towards offering the farmers a platform to aggregate, grade and sell their produce. This might also dissuade much of the illegal and informal trade. Investments to provide accurate information to farmers and collectors on herbal prices are also needed. Catalyst programmes, which federate farmers and collectors and generate volumes with buyback guarantees from processing companies, have been suggested as the means to move forward. These steps, along with greater investments in the extension services, drawing from traditional knowledge systems as well as modern technology would be critical steps that the Uttarakhand herbal sector could benefit from.

---

6. For a detailed overview of the value chain, see Annex V.

6.2.4. Seeds

Seeking to realise its potential as a model "Seed State", as outlined in the state's Agriculture Vision document, Uttarakhand has taken initiatives to promote the cultivation of seeds of cereals, pulses and a number of fruits and vegetables. The Terai Seeds and Development Corporation (TDC), Pantnagar, was established in 1968 as the pioneer research and delivery organisation for high yielding hybrid seeds in the country, and to date is the largest state seed corporation in the country. However, given its traditional focus on crops grown in the Indo-Gangetic plains to serve the larger interests of undivided Uttar Pradesh, the Corporation currently lacks a focus on the specific issues in hill agriculture. This is a problem the TDC is aware of and is seeking to correct over the years to come.

There are a number of infrastructural and institutional factors which prevent the expansion and growth of this sector. It seems additional efforts need to be made to promote seed production according to agro-climatic and topographic conditions. The TDC as well as research universities must be entrusted with the responsibility of addressing the seeds requirements of the 11 hill districts, and their capacity be built accordingly.

Vegetable seed production seems currently not very lucrative. This is because most vegetables produced in Uttarakhand receive price premiums owing to their seasonal advantage, and hence farmers have a great propensity to sell these vegetables in the market rather than retain them for seeds, or exclusively cultivate for seeds.

Seed production of a given crop is far more input-intensive than the crop itself. This is because seed production requires longer gestation, greater protection from the forces of nature, higher moisture, more intensive weeding and very labour-intensive processes of isolation, drying etc. Hence, even if support prices are offered for a crop, they may not well compensate the seed producing farmer for all the higher costs that seed production implies. Hence, there is much hesitation on part of farmers to take up seed production.

There are shortages in the public agricultural extension system in the state in general, and seeds extension is almost negligible. It seems that farmers need to be regularly educated on seed varieties, suitability, package of practices for particular varieties, land preparation, pre-harvest, harvest and post-harvest processes. Integrated pest and nutrition management also has to be emphasised. There is much need for targeted interventions to address these requirements. Area-specific initiatives, either through NGOs or cooperatives linked to potential buyers (possibly through contracts) could emerge as an effective means to promote local seeds development and harness Uttarakhand's potential to become a seeds provider for the whole country, including hilly areas. The private sector would have to play an important role in realising this potential, and the state could aid such efforts by promoting joint ventures in processing, marketing and distribution.

6.2.5. Floriculture

Floriculture is a high-growth industry worldwide. The international trade in floriculture is large and estimated to grow to US$ 16 billion by the year 2010 from a level of US$ 11 billion in 2006. These high growth rates at the international level in this sector also spill over in India. Although the value of exports of floriculture products from India has shown very significant growth, i.e., from US$ 0.5 million in 1993-94 to US$ 76 million in 2005-06, India is still a small player in the world floriculture trade. Key issues that are to be addressed in the Indian context include economies of scale, product range, year-round exports, quality controls and certification and cold chain management. The national market for flowers has also shown significant growth in recent years.

In Uttarakhand, it is estimated that the production of flowers amounted to 620 hectares in 2006-2007. The most popular varieties are rose, lilium, gerbera, calla lily, fresia, tulips, carnations and orchids. There seems quite some untapped potential in the production of flowers in the state given that flowers can be produced during the off-season period when production of these flowers seem difficult in the Indo-Gangetic plains. Flowers can be grown for nearly 11 months per year in Uttarakhand. Furthermore, the quality and flower production is better in hills than in plains, as cold and humid conditions are conducive for deep colour development of flowers, which are not possible in the plains (Agro-vision, 2010).

Under the assistance given by APEDA, a number of floriculture projects have been set up, many of which target niche export markets. Further floriculture parks are proposed to be established in the state with private sector investment. These would have common infrastructure facilities for production of flowers (including green houses, poly houses etc.), multiplication of planting material, sorting, grading, pre-cooling, cold storage, processing, packing and marketing. With assistance form a number of government agencies and banking institutions the state hopes to attract investments in export testing laboratories, flower auction centres, specific floriculture projects and a modern floriculture centre.

In order to improve the status of floriculture in the state, a number of measures need to be undertaken. The area under floriculture, currently at 620 hectares, has great potential to be expanded to as much as 1500 hectares according to the government of Uttarakhand. Much hand-holding may still be needed for making inroads into the export markets, which place much emphasis on quality and productions standards and processes; often ensured by high-grade planting material. This is a very specialised area and calls for technical support. Partnerships such as the Indo-Dutch Project are particularly valuable in this regard as they facilitate flow of new technologies and production processes into the state whilst also creating a stake for the local partners as well as foreign collaborators to continue the association. Projects built on these lines would require a strong extension component and effective demonstrations.

6.3. Simulations on the Pay-offs for Interventions

As to evaluate where interventions could best be done to improve value chains, we rely on simulations from data that reflect well the structure of the value chain. To start this exercise, we compare the cost structure of the different value chains of representative products of the fruit and vegetables value chains, i.e., apples, potatoes and tomatoes in Table 6.3. Given the high difference between high and low quality apples, we present separate calculations for these.[7] We find the highest marketing costs in the high quality apple value chain (Rs. 16/kg or 47 per cent of the retail price) and the lowest in the potato value chain (Rs. 5.3/kg or 32 per cent of the retail price). The level of marketing costs for tomato is in between (Rs. 7.4/kg). However, the marketing costs represent in this case a large share of the final retail price (57 per cent).

We further distinguish within the marketing costs the share of packaging costs, transportation costs, wastage costs at producers' and retailers' levels, commission rates (and other wholesale market costs) charged by the commission agents to the farmers and the retailers (based on the

7. The numbers in Table 6.3 were obtained through detailed interviews with farmers, wholesalers and retailers over the course of the project. While the results of these surveys are not based upon a representative scientific survey, we believe that they are however largely indicative of the situation on the ground.

Table 6.3 Prices and Values in the Apple, Tomato and Potato Commodity Chain in Uttarakhand

	Unit	Apple High quality	Apple Low quality	Potato	Tomato
Producer					
Quantities produced in Uttarakhand	tonnes	44928	67392	450637	76993
Producer price	Rs/kg	18.16	4.00	11.48	5.64
Total value production	million Rs	816	270	5173	434
Marketing sector					
Wholesale price	Rs/kg	24.04	8.12	14.04	7.55
Total marketing costs	Rs/kg	16.35	10.14	5.34	7.41
Consumer					
Quantities consumed	tonnes	38189	57283	428105.15	61594
Retail price	Rs/kg	34.51	14.14	16.82	13.05
Total value consumption	million Rs	1318	809	7200	804

Source: Own calculations.

survey detailed in Chapter 5), and the retail margin (Figure 6.3). In all cases, we look at the cost structure of a simple commodity chain where farmers deliver their produce to the *mandi* (or commission agents arrange for it to be picked up), commission agents then conduct an auction at the *mandi* and retailers purchase the produce there at that auction. Obviously, other longer value chains exist where consolidators sell to commission agents; where commission agents sell to commission agents or wholesale traders in other states; or where wholesalers buy at the auction, grade the produce and sell locally to retailers. However, the chosen structure represents the most common marketing chain.

In all cases, the retail margin—the margin between the wholesale price

Figure 6.3 Structure Marketing Margin between Farmers and Consumers

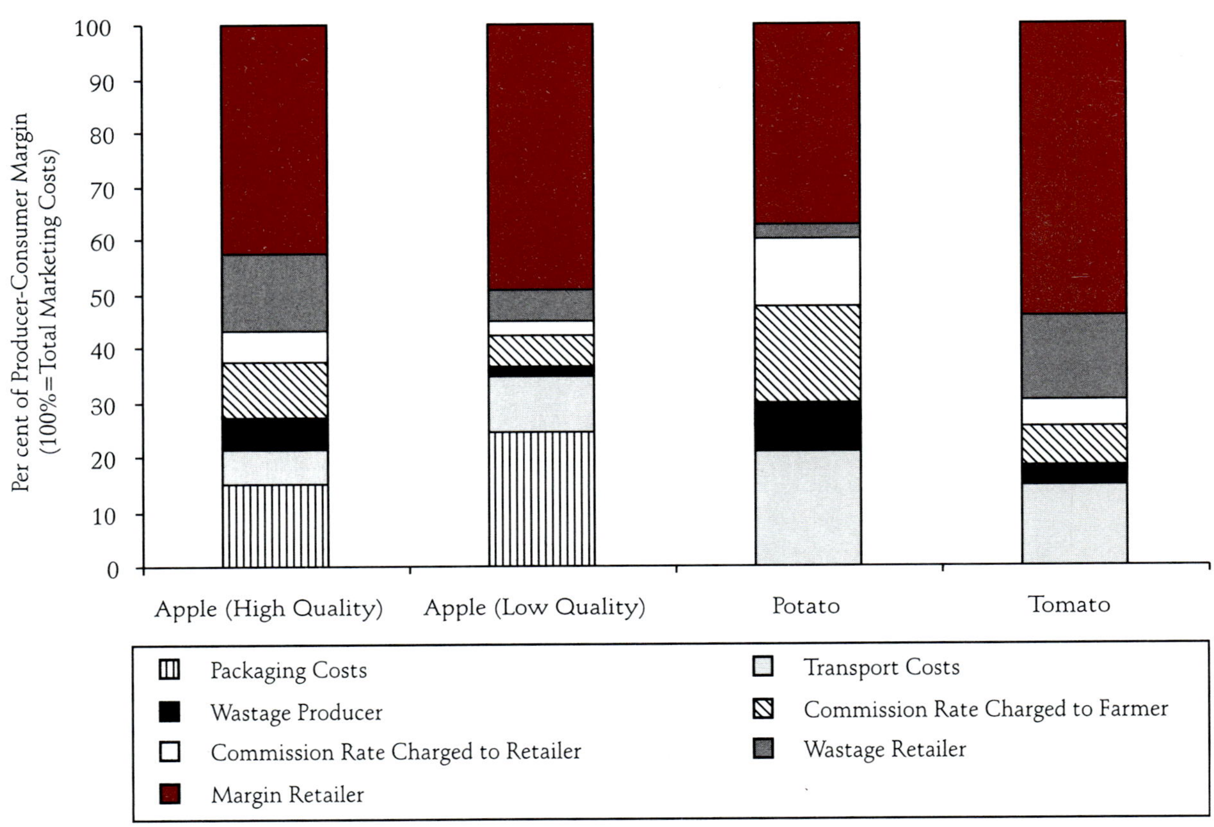

Source: Own calculations based on interviews with stakeholders.

and the price paid by the consumer—is the largest contributor to the marketing margin, 54 per cent, 47 per cent, 40 per cent, and 36 per cent in the case of tomato, low quality apple, high quality apple and potato respectively. The commission rates (plus other wholesale market costs) charged by commission agents make up 8 per cent and 16 per cent for low and high value apple, 12 per cent for tomato and 30 per cent for potato. Given the high value of high quality apples, the commission rate per kg of product are highest. This explains why apple—or fruit in general—commission agents are usually regarded as the ones who are better off. The highest wastage level is found for the most perishable crop, tomato, amounting to 20 per cent of the quantity marketed by the producer. The lowest level exists for the least perishable potato, 5 per cent.[8]

Table 6.4 Impact of Value Chain Interventions

Interventions	Crop			
	Apple		Potato	Tomato
	High quality	Low quality		
Total impact on consumer and producer surplus - in millions of Rs.				
Reduce taxes and commission rates (taxes=0%; commission rate=3%)	82	42	517	42
Reduce wastage costs reduction by 50%	71	33	133	50
Reduce transportation costs reduction by 50%	20	30	243	35
Reduction retail margin reduction by 50%	139	151	434	131
Eliminate packing costs reduction to 0%	99	151	-	-
All previous interventions combined	430	434	1347	273
Impact on producer price (in percentage)				
Reduce taxes and commission rates (taxes=0%; commission rate=3%)	+2.5	+3.8	+2.2	+2.5
Reduce wastage costs reduction by 50%	+2.1	+2.9	+0.6	+2.8
Reduce transportation costs reduction by 50%	+0.6	+2.7	+1.1	+2.1
Reduction retail margin reduction by 50%	+4.2	+13.6	+1.9	+7.7
Eliminate packing costs reduction to 0%	+3.0	+13.6	-	-
All previous interventions combined	+12.3	+36.6	+5.8	+15.1
Impact on consumer price (in percentage)				
Reduce taxes and commission rates (taxes=0%; commission rate=3%)	-4.6	-3.7	-5.5	-3.8
Reduce wastage costs reduction by 50%	-3.9	-2.9	-1.4	-4.4
Reduce transportation costs reduction by 50%	-1.1	-2.7	-2.6	-3.2
Reduction retail margin reduction by 50%	-7.9	-13.5	-4.6	-12.0
Eliminate packing costs reduction to 0%	-5.6	-13.5	-	-
All previous interventions combined	-23.1	-36.3	-14.2	-23.5

Source: Own simulations.

To evaluate the impact of interventions to reduce the costs in the value chains of the most important fruits and vegetables grown in the state, we compare different policies: a reduction of taxes and commission rates (taxes to 0 per cent and commission rates to 3 per cent), a reduction of wastage costs levels by 50 per cent, a reduction of transportation costs by 50 per cent, a reduction of the retail margin by 50 per cent, and an elimination of packaging costs in the case of apple. We value the impact of these interventions through a measure of the combined value of consumer and producer surplus expressed in millions of rupees (Table 6.4). These values indicate how consumers and producers would benefit from improvements in the value chain through higher (lower) prices received (paid) by all the producers (consumers) and through more traded quantities. We also present the effects on producer and consumer prices.

8. It is noteworthy that the largest part of the waste is reported by retailers when they open boxes and bags after purchase. It is not unlikely that this was bad produce put in there by producers.

While the evaluations are only approximate, and while we have no costs at our disposal to perform a systematic cost-benefit analysis, these are still indicative for setting policy priorities.[9]

Different important results emerge. First, because of the importance of potatoes in high-value agricultural crops in Uttarakhand, any improvement in the cost structure of the value chain outperforms almost any other interventions for all other crops.[10, 11] Second, due to the different structure of the value chain of each crop, different types of interventions might have a different priority in each value chain. For example, potato is a low retail margin crop and the biggest beneficial effect is obtained through a reduction of taxes and commission rates. Apple and tomato have relatively higher retail margins and we find the relatively largest effect through a reduction of the retail margin. Third, a reduction of transportation costs in the case of high-value agriculture has surprisingly little effect. This is because transportation costs make up only a small portion of the final retail price for this type of crop. Fourth, while a reduction of the wastage costs level would improve the situation for producers and consumers alike, the reduction in wastage costs levels does not show up as the most important intervention. Fifth, it is interesting to find that large effects in the value chain would be obtained by interventions in the regulatory environment. The elimination of taxes and de-licencing of agricultural marketing—inducing investments by modern retail, which would potentially reduce the retail margin (see below)—would have large pay-offs and would have relatively limited budgetary implications for the state government (the income from the mandi taxes represents 0.3 per cent of the expenditure budget of the state in 2006-07).

Our research on the impact of modern retail indicates that the conditions at the farmer level improve and that the prices that are charged to the consumers are at the lower end compared to traditional outlets. At the producer level, modern retail is offering better purchase conditions and higher prices. While our evidence in the state only concerns Mother Dairy (MD), as it is the only modern retailer that procures in the state, similar results emerge from experiences from other states and other modern retailers (see Box 6.2 on Reliance and its procurement of tomatoes in Rajasthan).

We document in Annex III in detail the reported advantages and disadvantages of farmers selling tomatoes to Mother Dairy (MD) as was found in a case study that we did in the Naugoan area in the district of Uttarakhasi. Farmers state various benefits from selling to MD: 1/ The pricing mechanism is transparent. 2/ Producer prices are higher. 3/ Prices are the same for everybody. 4/ There is no cheating with weights. 5/ Payments are regular. 6/ Bank payments avoid intra-household allocation problems. However, there are also several disadvantages. They include: 1/ MD only purchases the good quality produce. 2/ MD does not provide input advances. 3/ There is only a limited quantity that can be sold to MD.

---

9. Margins are fixed and additive. The effect of wastage is expressed as part of the margin between producer and consumer. Supply elasticities are assumed 0.20 and the demand elasticities are evaluated at 0.72. Mittal estimates demand elasticities for the fruit and vegetables group as a whole at 0.72 based on data of the NSSO in Mittal (2006). *Structural shifts in the demand for food: Projections for 2020*, ICRIER.

10. It is important to note that these simulations are based on the existing production structure and would fail to capture the impact of the introduction of new products given the dearth of information on the structure of their value chain.

11. These simulations only look at the interventions that affect the cost structure in the value chain and their welfare effect on consumers and producers. However, welfare could also be affected in other ways by changing value chains. For example, expanding value chains might have effects on input markets as well as employments effects. The impact of these changes are beyond the scope of this report given the lack of reliable data but should ideally be taken into account.

Modern retail might also be beneficial at the consumer level. Retail prices of potato and tomato were observed in a sample of modern retail (including Mother Dairy) as well as traditional outlets in different parts of New Delhi. The three graphs below (Figures 6.4, 6.5 and 6.6) show the results of this survey. While there is a significant overlap of prices in modern and traditional outlets, the prices that were observed in modern retail were all on the lower end compared to traditional retail. Additionally, quality was also often better in modern retail. Through better supply management, a reduction in wastage and going around licencing rents, it thus seems that modern retail is able to benefit producers and consumers alike, as predicted in our simulations.

Box 6.2 **The Advantages and Disadvantages of Selling to Modern Retail: The Case of Tomatoes and Reliance in Rajasthan**

There are several conveniences for farmers to sell to the collection centers (CC) of Reliance: 1/ Farmers save time in the transaction. Famers estimate that they save 3 to 4 hours in every sales transaction in CC compared to the regular *mandi*. The time needed for a transaction between leaving home and getting back is between 1 and 3 hours in CC and between 4 and 7 hours in *mandi* (where they have to wait for auction, traders, clients, better price, etc.); 2/ Transport and off-loading costs. When farmers sell to CC, they pay Rs. 5 for a bag of 50 Kgs, or Re. 0.10 /Kg. This compares to slightly higher costs in the *mandi*: Rs. 10 for a bag of 50 kgs (Re. 0.20 /Kg) as well as Rs. 2 to Rs.4 /Kg for off-loading for a bag of 50 Kgs. 3/ Farmers are assured by the correct weighing done in the CC. Electronic scales are being used by the CC. Only few commission agents on the *mandis* are using electronic scales. Most are mechanical scales in which farmers have less trust, especially when commission agents lack the appropriate weights. 4/ CC pays cash directly. In the *mandis*, this might or might not be the case. Some commission agents pay immediately but some farmers also have to wait a couple of days or a week before they receive the money.

Figure 6.4 **Tomato Retail Prices in New Delhi**

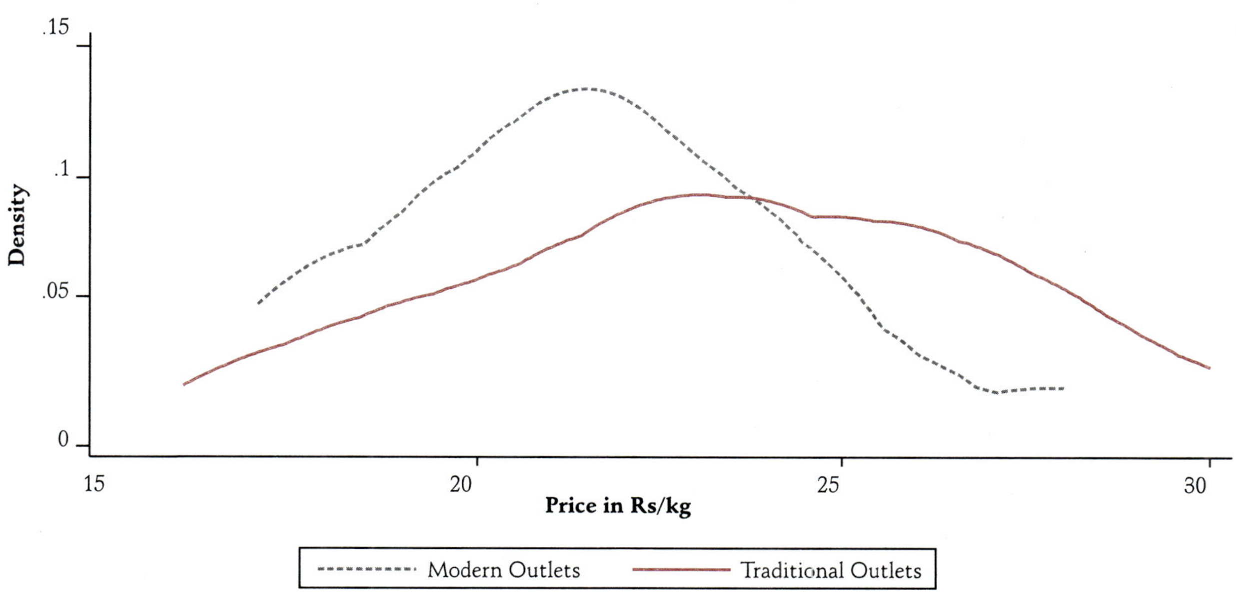

Source: Own retail survey.

Figure 6.5 Potato (Hill) Retail Prices in New Delhi

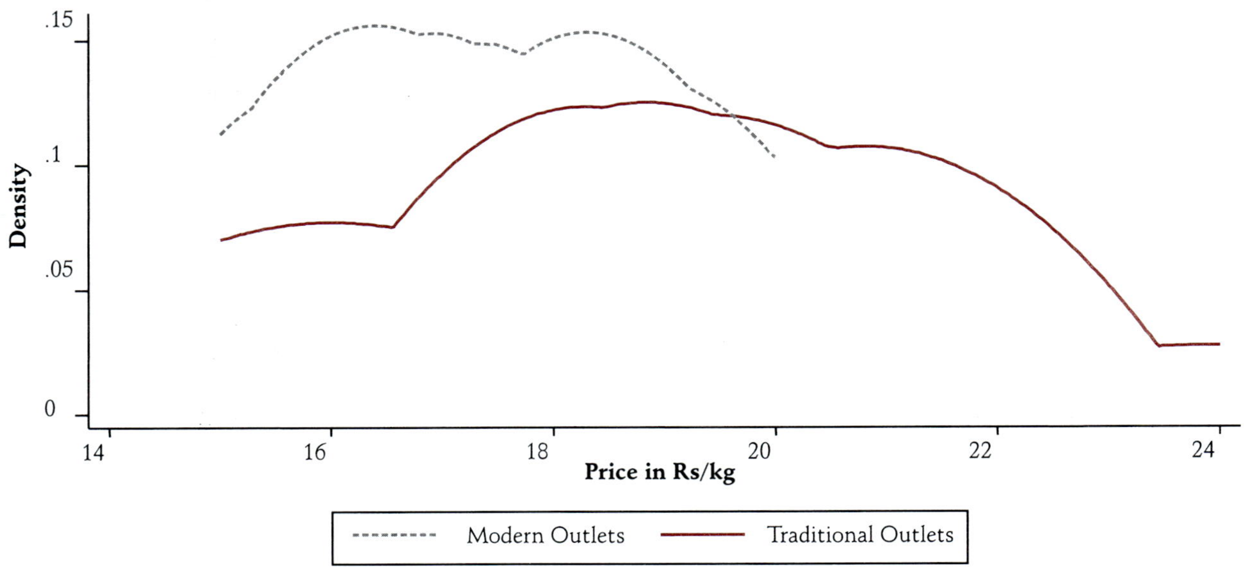

Source: Own retail survey.

Figure 6.6 Potato (Cold Storage) Retail Prices in New Delhi

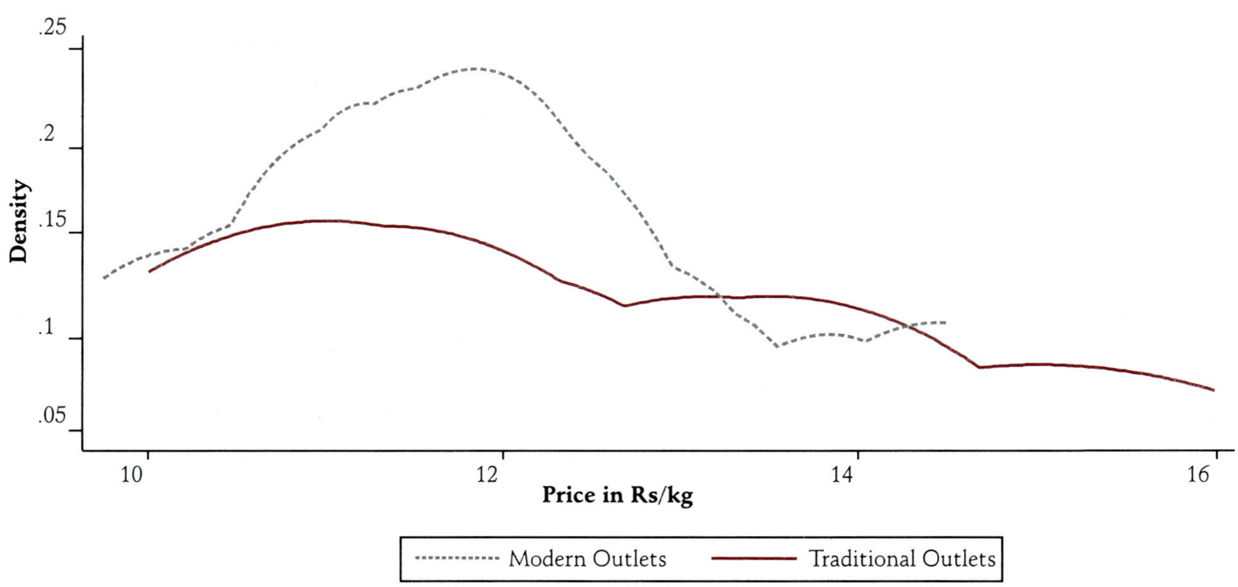

Source: Own retail survey.

7 The Way Forward

To better connect farmers to the market and to enable farmers to take advantages of the new opportunities offered in the market place, the state of Uttarakhand should change its role from tax collection and regulation to facilitation where a private market-driven agricultural environment is created that will benefit producers and consumers alike. To achieve the highest benefit for farmers as well as consumers, they should both be given possibilities of choice in market outlets. The state should turn attention to getting the institutions and infrastructure in place as to allow private trade to flourish.

To promote inclusive high-value agricultural growth, the state should focus, in the following sequence, on first, improving the competitive environment; second, providing information and relevant research; and third, upgrading infrastructure and modifying the financial sector. Differential strategies for agricultural development for the three regions (high mountain, mid-mountain, plains) in Uttarakhand are needed due to different agro-ecological characteristics and widely different accessibility to markets for these different regions. While some constraints (especially the institutional and regulatory issues) hold at the state level, investments should be tailored to the different zones of the state.

In Tables 7.1 and 7.2, we describe the current and desired situation for high-value agriculture in the short-term and long-term. The objectives for the agricultural sector would be to move from a weak market orientation to a highly diversified agriculture where off-seasonal vegetables, fruits, flowers, dairy and organic crops make up an increasing and important part of farmers' income. This strategy is sensible in the plains and the low mountains. Due to the geographical location and remoteness, prospects for agricultural commercialisation are however weak in the high-mountains. A strategy of promotion of off-farm activities, adoption of better food security crops and cultivation of crops which can only be grown in that area (mostly herbs) seems most appropriate to improve the welfare of the population in that area.

Step 1: Improve the Competitive Environment

Improving the competitive environment can be achieved through three measures: a change in market legislation; embrace better supply management; and an improvement of the fiscal structure and investment climate (Table 7.1).

a. Market legislation. Interviews with people involved in agricultural trade have over and over again articulated the problems with transparency, lack of competition and exploitation of the farmers. For example, it is difficult to justify the high margins of the commission agents given the limited number of services that they provide (i.e. all price risks and transportation losses before the auction are borne by the farmer). Different stakeholders are well aware of the problems of

Table 7.1 **Current and Desired Situation in High-value Agriculture—Required Policy and Institutional Changes**

	Current situation	Desired	
		Short-term	Medium-term/long-term
Objectives for crop growth			
High mountain	Weak market orientation Subsistence crops	Promote off-farm activities (tourism) Promote better food security crops High-value storable crops (MAP)	Promote off-farm activities (tourism) Promote better food security crops High-value storable crops (MAP)
Mid mountain	Weak market orientation Subsistence crops	Off-seasonal vegetables, fruits, flowers, dairy, organic	Off-seasonal vegetables, fruits, flowers, dairy, organic
Plains	Traditional staple production	Diversification strategy	Diversification strategy
Required changes in policies/regulations/procedures			
Marketing	Highly regulated through APM Act Complaints of high costs, because of licensing scheme Tax-collection driven APMC little market investments	Amend APMC Act Create competitive environment by allowing modern retail or modern supply management Improve *mandi* management Increase transparency	
Investment climate	Overlapping schemes Long delays before approval Paperwork burden	Greater coordination Reduce bureaucratic hurdles	
Fiscal incentives	Biased towards export crops	Remove bias	
Required institutional changes			
Research and extension	Extension not vey effective Research technology-focused Little post-harvest information Little research org. farming	Improve extension service Research post-harvest Research org. farming	Improve extension service Research post-harvest Research org. farming
ICT	Limited access to ICT and information	Improve access Strenghten Market Information Systems	
Contract farming Adherence to standards	Not allowed Weak capacity to meet market standards and tightening SPS standards	Amend APMC Act Develop marketing extension programs Research post-harvest Support capacity building with farmers and private sector for better compliance	 Implement marketing extension programs
Information-driven policy setting	Little updated and useful information Little monitoring/evaluation (M&E)	Put a system of regular representative household surveys in place useful for M&E	Continue system of household surveys

the current marketing system and they all expect that the situation is likely to improve when the APMC Act is amended.

Current marketing is regulated by the APMC Act. For each market area, a Marketing Committee is responsible for enforcing the Act and it is empowered to establish markets, control and regulate admissions to the market; charge fees (market, license and rental fees); issue and renew licences, and suspend or cancel them. The APMC Act in its current form adversely affects farmers and the agricultural sector as it restricts the choice of farmers. Farmers are required to sell their produce at the regulated markets and prevent them from selling to other channels that might offer better returns. Although the APMC Act states that it is the buyer that pays for the market fees and the commission rates, this is not the case in practice and even if this would be the case, these costs are, eventually, partly passed on to the farmer in terms of a lower farmgate price. The Act constraints the development and the modernisation of markets, by restricting these functions to the public sector. While the APMC collects significant revenues from market fees, the infrastructure on most markets is largely deficient with lack of water provision, covered areas, drainage and appropriate waste disposal. The existing Act however prevents greater private sector involvement in the construction and operation of wholesale markets. The amended Act proposes to remove the restriction of farmer direct marketing, opens market infrastructure development to other agencies and establishes a framework for contract farming.

While an amended Act would be a good step forward and would improve the marketing environment, it is however not adapted to a modern marketing system where competition between private players, food safety and vertical integration are enhanced. Even under the amended Act, a licensing system is still the norm and private investors and organised retail have to obtain licensing that could at any moment be revoked by a powerful committee where the interests of consumers and producers are underrepresented. Though such decisions can be challenged in courts of law, the costs for doing so are high. In general, the powers of the committee might dissuade investments. The proposed set-up hinders private investments and may encourage corruption. Under the amended Act, the emphasis is also still on regulation. For example, while contract farming would be allowed under the new Act, contracts have to be registered with the Committee and it has ultimate control on this. National and international experiences have shown that these are complications that might actually hinder contract farming. As shown from examples from other countries, it seems that less—not more—legislation might be a big boon for the farmer (see Box 7.1 for cases in Uttarakhand).

The marketing situation is not likely to change dramatically under the amended Act. The current APMC Act is relatively little respected and so might be the amended APMC Act. Effective commission rates are different than the ones prescribed by the APMC Act, contract farming—not allowed in the APMC Act—is already happening in some areas and illegal trade (i.e. produce that is not marketed on the *mandi*) is widespread (some informed sources put this at half the quantities marketed). Yet, in general it is hoped that under the provisions of the proposed legislation, the monopoly of existing *mandi* operators will be undermined and competition from more private players will be infused into the marketing system. More drastic measures are however needed to ensure healthy competition in the agricultural marketing system.

b. Better supply management. There are no unique models that connect small farmers better to the market. It seems that significant benefits for the farmers might be achieved through activities by market-oriented NGOs or through private sector investments. In India, lots of innovative value chain models are currently coming up, through NGO participation, cooperatives, private modern retailers as well cooperative retailers. Organised retail, in whatever form, constitutes an enormous opportunity for modernising the agricultural sector through its backward linkages and the services that it provides in input and credit delivery, in efficient supply management and in its better knowledge of the preferences of consumers. The current unorganised marketing sector, characterised by high costs, high wastage and a large number of intermediaries, is unable to deliver these services.

Several NGOs (such as HARC and CHIRAG) that are active in Uttarakhand are highly successful in better connecting farmers to the market. Different lessons emerge from their trials. It seems

Box 7.1 The Potential of Contract Farming in Uttarakhand

Several firms were asked about their views on contract farming during our field visits. Flexfood Foods Ltd. is one of the largest exporters of processed agricultural products in the state, especially processed mushrooms and culinary herbs. The stringent requirements imposed by the international buyers (it has certification from Europgap, ISO, and the British Retail Consortium) have made it obligatory for the firm to turn contract farming agreements. The firm provides seeds and other inputs free and thereafter deducts the costs of these over three installments of payments made for the produce. Over the period of cultivation, the farmers are subject to constant monitoring and supervision. Field surveyors visit plots at least twice a week to maintain the highest standards of quality. Prices are fixed before cultivation. The firm has a drop-out of 4 to 5 per cent of their farmers every year. The firm feels that there is no need for regulation in contract farming.

Another firm, Delicia is engaged in the processing and canning of fruits and vegetables. Their produce is sold to local companies such as Dabur or is used for exports. It buys its primary material through commission agents from different *mandis* in the state. They state that contract farming is impossible to implement for their business as farmers would side-sell their produce if prices on the *mandi* were higher than the agreed upon price while they would have an incentive to buy on the *mandi* if prices there were lower than an agreed upon price. They would thus not engage in contract farming even if the regulatory environment would be improved.

It thus seems that contract farming will only work in these situations where buyers have a monopsony position where products are required with specific characteristics that command a premium in the market and where the buyer is able to transfer some of these premiums to the producer (such as produce for export markets or products where there is a monopsony buyer as in the case of sugar cane and cotton). It seems that contract farming will thus only involve a small portion of the farmers.

important to conduct in-depth market research and establish links with potential buyers before encouraging farmers to grow specific crops. Their success is also often based on cluster based growing of the same crops (as to achieve economies of scale) and on significant planning of growing periods as to avoid gluts of specific produce. This can only be achieved through successful group formation.

Better supply management can also be achieved through increased investments by the private sector. Modern retail is eager to invest in Uttarakhand. While these investments might have enormous beneficial impacts on consumers and producers alike, modern retail has however until now not received a warm welcome in the state.[1] Interestingly, private rural business hubs are being rolled out in different parts of India where various services and products are retailed to rural households. It is estimated that the catchment area of the top four companies in India will be approximately 6 million farmers by the end of 2010.[2] These rural business hubs can potentially increase farmer incomes by improving output procurement, improving the quality of farm inputs available to the farmers, improving financial services available to farmers as well as retail at low costs. Attracting these investments in Uttarakhand might be of enormous help for the small farmer.

1. For example, Reliance Retail Ltd. has been refused—despite multiple requests—trader licenses in the largest *mandi* of the state (the Haldwani *mandi*) as vested interests clearly prefer the *status quo*.

2. The major players in this area are currently: DSCL (Hariyali Kisaan Bazaar), ITC (Choupal Saagar), Future Ventures India and Godrej Agrovet in joint venture (Aadhaar Retailing), Tata (Tata Kisaan Bazaar), and Reliance (Ranger Farms, and future Reliance Rural Business Hubs) as full or nearly full service hubs, extant or planned, and Mahindra (Mahindra Krishi Vihar) and Triveni with current focus on input provision.

A welcoming environment for the private sector could be created by offering them permission to purchase agricultural and horticultural products directly from farmers, by facilitating their access to land, and by promoting partnerships with local agricultural universities. The pay-offs would be large as alternative marketing outlets for farmers would create welcome healthy competition for the traditional marketing system.

c. Investment climate and fiscal incentives. Current value chains of major crops such as potatoes, tomatoes and apples are taxed while export crops, organic crops, medicinal herbs, etc., receive investment subsidies. These former crops are often grown by an important number of smallholders. It is unclear why smallholders should pay taxes while subsidies are simultaneously given for large scale investments in niche markets, usually for exports. It is also an important policy question on who benefits from such an implicit transfer of funds. In any case, a common misperception is that export markets are highly lucrative but is often forgotten that they are also often costly to conquer.[3]

Local food demand in India is changing fast and new opportunities are emerging rapidly. The state of Uttarakhand has important advantages with respect to other Indian states to exploit these. They include closeness to important urban centers, seasonal advantages, and diversity of climate. Investments should also be geared to best use these advantages in local markets as these are often easier to compete in, rather than distant and often impenetrable export markets.

Ministries often offer grants to the private sector to encourage greater investments. Investors complained about the paperwork needed, the lack of transparency as well as the delay in approvals of subsidies. The investment climate would further benefit significantly from better coordination of these schemes. For example, three different ministries offer investment grants for cold storage development with different terms and conditions (World Bank, 2007).

Step 2: Provide Information and Relevant Research

After the improvement of the competitive environment, farmers would have incentives to upgrade to better production and marketing technologies. To allow them to do so, stakeholders will need access to relevant agricultural information.

a. Market and agricultural information. Public information on agriculture and agricultural markets is currently weak. The needed information should be reliable, timely, publicly available and relevant. First, imprecise statistics make planning and investment hard. For example, different sources on apple cultivation show widely varying acreage numbers, ranging between 3,500 hectares and 55,000 hectares. Second, statistics are outdated. The results of the agricultural census of 2001 are at the time of writing of the report not available. Investments are needed to speed up the process of access to information to allow policy-makers and stakeholders to react quickly to rapidly changing environments. Third, statistics are not publicly available. For example, time series of Agmarknet data—price and quantity data that are regularly collected at each market—are hard to access. The lack of easy access to these data makes market analysis cumbersome and investments more risky. Fourth, there is currently no information available on important information such as which type of farms grow which crops, the level of marketed surplus, marketing conditions, use of inputs, adoption of modern technologies, the location of agricultural production of specific crops,

3. For example, while Uttarakhand exported for US$300,000 of basmati rice, litchis and flowers (three Agri-Export Zone products) in 2005/06, the yearly budget of the agri-export unit alone is already US$100,000. On top of this, APEDA and NHB also provide investments subsidies towards exports. While exact spending is difficult to get at, it is quite likely that the government spends annually more on agricultural export promotion than the annual value of agricultural exports.

beneficiaries of government schemes, etc. It seems that a regular, representative and comprehensive farm survey at the state level would help policy-makers and stakeholders enormously in better understanding the current agricultural situation, in evaluating the effect of current policy design and thus in better anticipating the impact of policy changes.

Finally, information access for farmers would need attention. Farmer-friendly media such as telephones, village *panchayat* offices and internet hubs (where they exist) should be effectively leveraged to provide farmers better information on prices of various crops in the markets to improve the small farmers' bargaining positions. These types of investments are currently done by the Ministry of Agriculture where agricultural portals are developed that would allow easier linkages between farmers and extension agents as well as allow access to updated price and crop production information. It would be useful to do solid evaluation of these programmes as to potentially further scale it up. One would potentially also bring private players into this provision of information as to better assure its sustainability.[4]

b. Agricultural research and extension. Market-led research and extension models are required that are especially tailored for hill areas.[5] Knowledge creation at the local level is needed and capacity building requirements seem high. Priority setting in agricultural research could be better aligned with the production constraints that farmers are facing in the field. Some examples of our field visits illustrate some of these issues. For example, a recurrent problem stated by farmers as a major constraint in agricultural production is the destruction of agricultural produce by boars, monkeys or other wildlife. Little research or extension has been done on this important production problem.[6] Second, most of the agricultural research has over the years been oriented towards the development of better agricultural varieties for the plains (especially before the split-off of the state from Uttar Pradesh). This was a logical choice given the homogenous production conditions in the plains. Production conditions in the hills on the other hand are different and more heterogeneous. This makes research and extension more challenging but does not imply that it should not be done. Third, priority setting in agricultural research should be guided by economic pay-offs of potential findings. While research institutions that conduct research on 'herbal plants' are plenty, little is done, for instance on the most important vegetable of the state, potato, as well as on organic production practices.

The existing extension machinery in the state is widely reported to be outdated and lacking in delivery capacity for the needs of the state. This calls for regular and timely training and upgrading of extension officers' skills, making sufficient appointments for such positions (including the Uttarakhand Organic Commodities Board training and extension activities) and partnering with private extension workers as well as NGOs who may be better equipped to deliver the necessary services. Most of the extension in the state has further been geared towards production technologies and little has been done on post-harvest technologies.

It would be a sensible strategy of the government to invest in marketing extension programs where communities are trained in better understanding markets and in better marketing those products that the market demands. Specific training could be conducted by NGOs to help communities build capacity in this area. Successful experiences from other countries and other areas in India are

---

4. For example, the company Reuters is providing through SMS price information, weather information, and crop advisory services to more than 20,000 farmers in the state of Maharashtra for a fee of about Rs.2/day.

5. For an overview of issues with demand-driven extension in India, see Birmer and Anderson, 2007.

6. Some recent research shows for example that innovative bio-insecticides might potentially reduce this problem (as tried out by the NGO Inhere).

available that could be used as models for this type of programme (See for example, FAO, 2005).[7] Given the importance of women in agriculture in the state of Uttarakhand, special attention towards them is needed. It is thus crucial that specially tailored programmes are built in for them in these marketing extension programmes.

Step 3: Upgrade Infrastructure and Modify the Financial Sector

Better infrastructure will contribute significantly to better functioning input and output markets. However, while this is an important objective to aim for, these investments are costly and will take more time to be realised (Table 7.2).

a. Transport infrastructure. Given the scattered population and the mountainous terrain, transport infrastructure in the state of Uttarakhand is relatively costly to improve compared to the rest of India. The state government has made significant investments to improve road infrastructure in recent years and transport costs have very evidently come down considerably. There have also been a number of investments in ropeways to connect isolated villages to the road. These investments have large pay-offs on welfare, through agricultural as well as off-farm income linkages. For example, Fan *et al.* (2007) show the large returns on roads for agricultural growth and poverty alleviation in India. However, while better transport infrastructure will improve welfare of the population by better access to markets—as well as to social services—the effect of transport infrastructure on high-value agriculture specifically is relatively limited as shown above.

b. Irrigation infrastructure. Lack of irrigation water seems especially problematic in the mid- and high-mountains. In order to provide irrigation in the rain fed areas, rainwater harvesting has to be developed and small and medium sized reservoirs for the collection of rainwater should increasingly be built. Especially sprinklers and drip irrigation can be promoted to encourage the efficient use of water (Kar, 2006).

c. Marketing infrastructure. Market infrastructure should be improved. An amount of Rs.331 million (US$ 8 million or 1.35 times the budgetary allocation that is yearly spent on horticulture in the state) was collected as market fees in the state of Uttarakhand in the year 2006-07 but it is unclear to what extent these revenues helped to improve the physical marketing environment for farmers. While it is stated that part of this revenue is used to towards investments for better marketing conditions, this is not always the case. Food safety is especially an important issue. Most of the markets have minimal sanitary regulations and in most markets, wastage lays around for days, animals are free to roam around on the market, and drainage systems are clogged following heavy rainfall. The food marketing infrastructure is not well adapted to the new challenges of food marketing, especially of food safety as increasingly demanded by consumers, locally as well as internationally. Major investments are needed to upgrade these traditional markets.

d. Financial sector. Lack of access to credit is often cited by farmers as a major constraint in production and marketing activities and they thus rely on informal credit, often tied to traders. It seems that farmers would benefit enormously by a broader adoption of innovative products such as group lending and *kisan* credit cards. Farmers involved in high-value agriculture face often also severe calamities, more so than in traditional agriculture. Crop insurance schemes could further be envisioned as to encourage investments in these riskier crops.

7. Marketing extensions programmes are characterised by activities to help farmers increase profitability. An increase in profitability can be achieved through the selection of profitable crops in that particular area, increasing sales by improving access to buyers and markets, reducing costs, and reducing post-harvest losses. Information gathering is needed to give farmers reliable advice on the farming area, the crops grown and the costs of producing them, and the market and those involved in supplying it. Gathering that information can be done through interviews with traders and farmers (FAO, 2005).

Table 7.2 Current and Desired Situation in High-value Agriculture—Required Infrastructure Investments and Financial Sector Modifications

	Current situation	Desired	
		Short-term	Medium-term/long-term
Infrastructure investments			
Irrigation	Good in plains	Infrastructure maintenance	Infrastructure maintenance
	Limited in mid-mountain and high-mountain	Water harvesting projects and drip irrigation	Water harvesting projects and drip irrigation
Transport	Relatively good in plains	Road maintenance	Road maintenance
	Limited in mid-mountain	Road maintenance/ropeways	Road maintenance/ropeways
	Problematic in high-mountain	Road maintenance/ropeways	Road maintenance/ropeways
Market infrastructure	*Mandi* in plains	Improve infrastructure	
	Limited in mid-mountain	Establish collection centers in areas where aggregation is promising	
	Limited in high-mountain		
Financial sector modifications			
Credit delivery	Limited access to formal credit Informal credit through traders	Support innovative credit approaches (group lending, *kisan* credit cards, etc.)	
Insurance scheme	No crop insurance	Promote crop insurance schemes	

References

Acharya, S.S. (2004). "Agricultural marketing in India", Vol. 17, *State of the Indian Farmer: A Millenium Study.* Government of India. New Delhi: Academic Foundation.

Agricultural Statistics Division (ASD) (2007). *Land Use Statistics in Uttarakhand at a Glance.* Dehradun: Government of Uttarakhand, Chief Revenue Commissioner.

Agri Export Advantage (2006). "Organic food industry: Trends and opportunities". *Agri Export Advantage* V(II) March. Export-Import Bank of India. (Article made available at *www.eximbankagro.com*)

Agri-Export Development Unit (AEDU) (2006). *Annual report: 2004-2005; 2005-2006.* Dehradun.

Alam, G. (2007). *Appraisal of Financial Mechanisms, Marketing Strategies and Value-Adding Opportunities in the Organic Sector as Tools for Enhancing Farmers' Income-Generation : A Study of Organic Basmati Rice in Uttarakhand, India.* Dehradun: Centre for Sustainable Development.

Alam, G. and D. Verma (2006). *Cost Benefit Analysis of Organically Produced Crops Grown in Dehradun and Udhamsingh Nagar.* Dehradun: Centre for Organic Farming.

Athukorala, P.C. and S. Jayasuriya (2003). "Food safety issues, trade and WTO rules: A developing country perspective", *World Economy* 26(9): 1395-1416.

Banerji, A. and J.V. Meenakski (2004). "Buyer collusion and efficiency of government intervention in wheat markets in northern India: An asymmetrical structural auction analysis", *American Journal of Agricultural Economics* 86(1): 236-253.

Bell, C. and T.N. Srinivasan (1989). "Interlinked transactions in rural markets: An empirical study of Andhra Pradesh, Bihar and Punjab", *Oxford Bulletin of Economics and Statistics* 51(1): 73-83.

Bhadani, K.N. (1998). "Entreprise-based transformation of hill agriculture: A case study of vegetable growing farmers in Garampani area, Nainital district, India", *Discussion paper*, Series no. MEI 98/5. Kathmandu, Nepal: Icimod.

Bigsten, A., P. Collier, S. Dercon, M. Fafchamps, B. Gauthier, J.W. Gunning, A. Isaksson, A. Oduro, R. Oostendorp, C. Patillo, M. Soderbom, F. Teal and A. Zeufack (2000). "Contract Flexibility and Dispute Resolution in African Manufacturing", *Journal of Development Studies* 36(4):1-37.

Birmer, R. and J. Anderson (2007). "How to make agricultural extension demand-driven? The case of India's agricultural extension policy", International Food Policy Research Institute, *Discussion Paper* no 729.

Birthal, P., P. Joshi and A. Gulati (2005). "Vertical coordination in high-value food commodities: Implications for smallholders", Technical report, IFPRI, *MTID Discussion Paper* 85, Washington DC.

CIP (2006). *World Potato Atlas, India.*

CITES (2002). "Convention on international trade in endangered species of wild fauna and flora". *www.cites.com*

Chandola, S. (2007). "About the medicinal plants of Uttaranchal in brief", in *Uttarakhand Herbal Expo 2007: A Souvenir.* Dehradun: Uttarakhand Forest Development Corporation.

Chauhan, N.C. (2006). "Scope of commercialization of medicinal and aromatic plants", Science Tech. *Earthquake*, July.

Crow, B. and K.A.S. Murshid (2003). "Economic returns to social power: Merchants' finance and interlinkage in the grain markets in Bangladesh", *World Development* 22(7): 1011-1030.

CTA (2002). *Medicinal Plants: Pluck, not Plunder.* Spore No. 99, June: 1 – 2 , 11.

Deodhar, S.Y., M. Landes and B. Krissoff (2006). *Prospects for India's Emerging Apple Market.* USDA, FTS-391-01.

Dercon, S. (2005). *Insurance against Poverty.* Oxford: Oxford University Press.

Dercon, S. and L. Christiaensen (2007). "Consumption risk, technology adoption, and poverty traps: Evidence from Ethiopia", World Bank Policy Research *Working Paper* No. 4257.

Deshingkar, P., U. Kulharni, L. Rao and S. Rao (2003). "Changing food systems in India: Resource sharing and marketing arrangements for vegetable production in Andhra Pradesh", *Development Policy Review* 21 (5-6):627-39.

Devadoss, S. and T. Wahl (2004). "Welfare impacts of Indian apple trade policies", *Applied Economics* 36(12).

Diop, N. and S.M. Jaffee (2005). "Fruits and vegetables: Global trade and competition in fresh and processed product markets", in (eds), M.A Aksoy and J.C. Beghin (eds.), *Global Agricultural Trade and Developing Countries.* World Bank. pp.237-257.

Down to Earth (2006). "Almora's dying waterbodies", *Down to Earth* 23(12).

Fafchamps, M. (2004). *Market Institutions in Sub-Saharan Africa*. Cambridge, Mass: MIT Press.

Fafchamps, M. and B. Minten (1999). "Relationships and traders in Madagascar", *Journal of Development Studies* 35(6): 1-35.

———. (2001). "Property Rights in a Flea Market Economy", *Economic Development and Cultural Change* 49(2): 229-268.

Fafchamps, M. and R.V. Hill (2005). "Selling at the farm-gate or traveling to market", *American Journal of Agricultural Economics* 87(3): 717-734.

Fafchamps, M., R. Vargas-Hill and B. Minten (2007). "Non-staple Food Markets in India", International Food Policy Research Institute, *Discussion Paper* no 717.

Fan, S., A. Gulati and S. Thorat (2007). "Investments, subsidies and pro-poor growth in rural India", *IFPRI discussion paper* 716. Washington DC.

FAO (2005). *Horticultural Marketing: Marketing Extension Guide*. Rome: FAO.

FRLHT (Foundation for Revitalization of Local Health Traditions). (1997) *Medicinal Plants of India: Guidelines for National Policy and Conservation Programmes*. Bangalore: FRLHT.

Gabre-Madhin, E. (2001). "Market institutions, transactions costs, and social capital in the Ethiopian grain market", IFPRI *Research Report* 124, Washington DC.

Goyal, A. (2008). "Information technology and rural markets: Theory and evidence from a unique intervention in Central India", University of Maryland. *(mimeo)*.

Government of Uttarakhand (GoUA). (2007). *New Series of Estimates of State Domestic Product of Uttarakhand: 1999-00 to 2005-06*. Directorate of Economics and Statistics, Government of Uttarakhand, Dehradun.

———. (2007a). *Land Use Statistics in Uttarakhand at a Glance*. Dehradun: Agriculture Statistics Division, Office of the Chief Revenue Commissioner, GoUA.

———. (2007b). *Statistical Diary: Uttarakhand 2005-06*. Dehradun: Directorate of Economics and Statistics, Planning Department, Government of Uttarakhand.

———. (n.d. a). *Investment Opportunities in Uttaranchal*. Dehradun: Forest and Rural Development Branch, Uttarakhand Secretariat, Government of Uttarakhand.

———. (n.d. b). *Development Initiatives in Rural Uttaranchal: enhancing Livelihood Opportunities with Sustainable Development*. Dehradun: Office of the Forest and Rural Development Commissioner, Government of Uttarakhand.

———. (n.d. c). *Food Processing Industries: Advantage Uttaranchal*. Ranikhet: Directorate of Horticulture, Government of Uttarakhand.

Gulati, A., Minot, N., Delgado, C. and Bora, S. (2007). "Growth in high-value agriculture in Asia and the emergence of vertical links with farmers", in J. Swinnen (ed.), *Global Supply Chains, Standards and the Poor*. Wallingford: CABI Publishing. pp. 91-108.

Gulati, A., T. Reardon, K. Ganguly and K. Gupta (2008). "Transforming agri-food systems: Role of organized retail in India", Keynote address as first *Annual Conference of the Indian Society of Agri-Business Management*. Bangalore.

Hsieh, C. and E. Moretti (2003). "Can free entry be inefficient? Fixed commissions and social waste in the real estate industry", *Journal of Political Economy* 111(5).

IFAD (2005). "Organic agriculture and poverty reduction in Asia: China and India focus", *Report no*. 1664.

ITCOT (Industrial and Technology Consultancy Organisation of Tamil Nadu Limited) (1999). *Report on Ayurvedic Medicines*. Chennai: ITCOT.

Jaffee, S.M. and S. Henson (2004). "Standards and agro-food exports from developing countries: Rebalancing the debate", World Bank Policy Research *Working Paper* 3348.

Jain, S.K. (1987). "Endangered species of medicinal herbs in India", *Medicinal Herbs in Indian Life* 16(1): 44–53. Quoted in: World Bank, tech. Series 355.

Joshi, A., P.S. Garia and I. Hasnain (1999). *Uttarakhand Update: A Statistical Profile for Development Planning*. New Delhi: Kanishka Publishers and Distributors.

Joshi, P.K., A. Gulati and R. Cummings (2007). *Agricultural Diversification and Smallholders in Asia*. New Delhi: Academic Foundation.

Joshi, P.K., P.S. Birthal and N. Minot (2006). "Sources of Agricultural Growth in India: Role of Diversification towards High-Value Crops", MTID *Discussion Paper* No. 98, IFPRI.

Kala, R.P. and R.S. Negi (2007). "Aushadhiya vanaspati evam van sansthaa, Uttaranchal" (in Hindi), in *Uttarakhand Herbal Expo 2007: A Souvenir*. Dehradun: Uttarakhand Forest Development Corporation.

Kar, S. (2006). *Inclusive Growth in Hilly Regions: Priorities for the Uttarakhand Economy*.

Katyal A. (2006). *Appraisal of Certification Costs for Farmers and Farmer's Organizations under Different Certification Schemes-Sunstar Overseas Limited, India*. Prepared for Food and Agricultural Organization, Rome.

Khatana, V. (2003). "Potato farming and innovation", *India Together*.

KIT (2003). "Cultivating a healthy enterprise: Developing a sustainable medicinal plant chain in Uttaranchal – India", *KIT Bulletin* 350, Amsterdam.

Kranton, R. (1996). "Reciprocal exchange: A self-sustaining system", *American Economic Review* 86(4): 830-851.

Krishna, S. (2002). "Economic profile of Uttarakhand", *Economic and Political Weekly*. May 11.

Kumari, B., V.K. Madan, J. Singh, S. Singh and T.S. Kathpal (2004). "Monitoring of pesticidal contamination of farmgate vegetables from Hisar", *Environmental Monitoring and Assessment* 90(1-3): 65-71.

Kumar, B. (2006). "Almora's drying waterbodies", *Down To Earth*. June 15.

Lal, S.D. (2003). "Apple farming in Uttaranchal: Where does it stand?", in K.K. Jindal, R. Bawa and T. Pratap (eds.), *Apple Farming and Livelihoods in the Himalayas*. Dehradun: Shiva Offset Press. pp.161-163.

Lele, U. (1971). *Food Grain Marketing in India*. Cornell University Press.

Mariental India Pvt Ltd. (2006). *Comprehensive Study on the Processing Of Apple*.

Marshall, F., R. Agarwal, D. te Lintelo, D.S. Bhupal, R.P.B. Singh, N. Mukherjee, C. Sen, N. poole, M. Agrawal and S.D. Singh (2003). "Heavy metal contamination of vegetables in Delhi". Report prepared for the UK Department for International Development. *(mimeo)*.

Mattoo, A., D. Mishra and A. Narain (2007). *From Competition at Home to Competing Abroad*. World Bank.

McMillan, J. (2002). *Reinventing the Bazaar: A Natural History of Markets*. New York: Norton. p.256.

McMillan, J. and C. Woodruff (1999). "Dispute prevention without courts in Vietnam", *Journal of Law, Economics and Organization* 15(3): 637-658.

MCX (2006). "Potato" *(mimeo)*.

Minot, N. and R. Hill (2007). *Developing and Connecting Markets for the Poor*. IFPRI, Policy brief.

Mishra, S. (2006). *Organic Drive in Uttaranchal Ignores Farmers' Concerns*. November 6.

Mittal, S. (2006). *Structural Shift in the Demand for Food: Projections for 2020. ICRIER Working Paper* no 184.

————. (2007). "Can horticulture be a success story for India?", *ICRIER Working Paper* 197

Murthy, K.S.R and G.K. Kadekodi (2000). "Dynamics of water resources in Kumaon region", in G K Kadekodi, K S R Murthy and Kireet Kumar (eds.) *Water in Kumaon: Ecology, Value and Rights*. Nainital: Gyanodaya Publishers.

Natawidjaja, R., T. Reardon. and S. Shetty, with T.I. Noor, T. Perdana, E. Rasmikayati, S. Bachri and R. Hernandez (2007). *Horticultural Producers and Supermarket Development in Indonesia*. UNPAD/MSU Report published by the World Bank/Indonesia.

North, D. (1990). *Institutions, Institutional Change, and Economic Performance*. New York: Cambridge University Press.

Olson, M. (1965). *The Logic of Collective Action*. Cambridge, MA: Harvard University Press.

Orden, D. and D. Roberts (2007). "Food regulation and trade under the WTO: Ten years in perspective", Proceedings of the *26th Conference of the International Association of Agricultural Economists*. Blackwell Publishing. pp.103-118.

Palaskas, T.B. and B. Harriss-White (1996). "The identification of market exogeneity and market dominance by tests instead of assumption: An application to Indian material", *Journal of International Development* 8(1): 11-23.

Planning Commission (2000). *Report of the Task Force on Conservation and Sustainable of Medicinal Plants*. New Delhi: Government of India Press.

Ratanpal and S.D. Singh (2007). "Nayi dishaa naye aayaam: Uttarakhand mein jadi-booti sangrahan evam vipanan" (in Hindi), in *Uttarakhand Herbal Expo 2007: A Souvenir.* Dehradun: Uttarakhand Forest Development Corporation.

Rathore, D.S. (2003). "Future of Apple farming in Uttaranchal: Some issues", in K.K. Jindal, R. Bawa and T. Pratap (eds.), *Apple Farming and Livelihoods in the Himalayas*. Dehradun: Shiva Offset Press. pp.169-182.

Ramakrishnappa, K. (2002). *Impact of Cultivation and Gathering of Medicinal Plants on Biodiversity: Case Studies from India*. FAO, Biodiversity and the Ecosystems Approach in Agriculture, Forestry and Fisheries. Interdepartmental Working Group on Biological Diversity for Food and Agriculture. Rome.

Ramaswami, B. and P. Balakrishnan (2002). "Food Prices and the efficiency of public intervention: The case of public distribution system in India", *Food Policy* 27(5-6): 419-36.

Rawat, M. and J. Mishra (2001). "An approach to sustainable development of agriculture in the UP Himalaya", in M. Rawat, R. Negi and M. Singh (eds.), *Land-use management and sustainable development in Uttaranchal Himalaya*. Srinagar: Transmedia. pp.176-186.

Reardon, T. and C. Barrett (2000). "Agroindustrialisation, Globalization and International Development: An Overview of Issues, Patterns and Determinants", *Agricultural Economics* 23:195-205.

Reardon, T. and J. Berdegué (2002). "The rapid rise of supermarkets in Latin America: Challenges and Opportunities for Development", *Development Policy Review* 20(4): 371-88.

Reardon, T. and A. Gulati (2008). "The rise of supermarkets and their development implications: International experience relevant for India", *IFPRI discussion paper.*

Reardon, T., C. Timmer, C. Barrett and J. Berdegue (2003). "The rise of supermarkets in Africa, Asia and Latin-America", *American Journal of Agricultural Economics* 85(5):1140-1146.

Reardon, T., A. Gulati and B. Minten (2008). "The rapid rise of supermarkets in India: Implications for farmers, processors and traders". *(mimeo)*.

Rosenzweig, M.R. and K.I. Wolpin (1993). "Credit constraints, consumption smoothing, and the accumulation of durable production assets in low-income countries: investments in bullocks in India", *Journal of Political Economy* 101: 223-244.

Sati, V.P. (2005). "Natural resource conditions and economic development in the Uttaranchal Himalaya, India", *Journal of Mountain Science* 2(4): 336-350.

Sawhney, A. (2005). "Quality Measures in Food Trade: The Indian Experience", *World Economy* 28(3): 329-348.

Sajwan, B.S. (2007). "Government initiatives for the development of medicinal plants sector in India: Role of the forest department", in *Uttarakhand Herbal Expo 2007: A Souvenir.* Dehradun: Uttarakhand Forest Development Corporation.

Sharma, R. (2007). "Aushadhiya evam sagandha paudhon ka vishwa bazaar evam niryaat sambhavanaayen" (in Hindi), in *Uttarakhand Herbal Expo 2007: A Souvenir.* Dehradun: Uttarakhand Forest Development Corporation.

Shiva, Vandana (2007). "Corporate hijack of retail: Retail dictatorship vs retail democracy." A Report by Navdanya/Research Foundation for Science, Technology and Ecology. New Delhi, India.

Singh, S. (2002). "Contracting out solutions: Political economy of contract farming in the Indian Punjab", *World Development* 30(9): 1621-38.

Singh, S.K. (2007). "Uttarakhand mein aushadhiya evam sagandha paadapon ki kheti naye paridrishya mein nayi neetiyon ka pratipadan" (in Hindi), in *Uttarakhand Herbal Expo 2007: A Souvenir.* Dehradun: Uttarakhand Forest Development Corporation.

Srivastva, A. (2006). *Appraisal of Certification Cost for Farmers and Farmers Organisations under Government Scheme in Uttaranchal.* Prepared for the Food and Agricultural Organization, Rome.

Storm, S. (1997). "Agriculture under trade policy reform: A quantitative assessment for India", *World Development* 25(3): 425-36.

Subrat, N., M. Iyer and R. Prasad (2002). *The Ayurvedic Medicine Industry: Current Status and Sustainability.* India Country Sub Study for International Collaborative Research Project: Instruments for Sustainable Private Sector Forestry. New Delhi: Ecotech India Pvt. Ltd. and London: International Institute for Environment and Development.

Swinnen, J. (2007). *Global Supply Chains, Standards and the Poor.* Wallingford: CABI Publishing.

Thomas, S. (n.d.). "Agricultural commodity markets in India: Policy issues for growth". Indira Gandhi Institute for Development Research. *(mimeo).*

Tolia, R.S. (2007). *Inside Uttarakhand Today.* Dehradun: Bishen Singh Mahendra Pal Singh Publishers.

Townsend, R.M. (1994). "Risk and insurance in village India", *Econometrica* 62: 539-591.

Umali-Deininger, D.L. and K. Deininger (2001). "Towards greater food security for India's poor: Balancing government intervention and private competition", *Agricultural Economics* 25(2-3): 321-35.

Umali-Deininger, D. and M. Sur, (2007). "Food Safety in a globalizing world: Opportunities and challenges for India", in Proceedings of the *26th Conference of the International Association of Agricultural Economists*. Blackwell Publishing. pp. 135-147.

UNCTAD (2004). *Trading Opportuinities for Organic Food Products From Developing Countries: Strengthening Research and Policy-making Capacity on Trade and Environment in Developing Countries.* New York and Geneva: United Nations Conference on Trade and Development.

Uttarakhand State Planning Department (2007). *Annual Plan 2006-07.* State Planning Department, Government of Uttarakhand. Made available at *http://gov.ua.nic.in/planning/annualplan0607/INDEX.htm*

Virdi (2004). "Wild Plants as resource: New opportunities or last resort? Some dimensions of the collection, cultivation and trade of medicinal plants in the Giro Basin", in G. Alam and J. Belt (eds.), *Searching Synergy: Stakeholder Views on Developing A Sustainable Medicinal Plant Chain in Uttaranchal–India.* KIT (Royal Tropical Institute) Bulletin 359, Amsterdam.

Vogtmann, H. (2004). *Sustainable Collection of Medicinal Plants: Towards International Standards and Criteria.* Bangkok, 19 November.

von Braun, J. and E. Kennedy (1994). *Agricultural Commercialization, Economic Development and Nutrition.* John Hopkins University Press.

World Bank (2005). "Food safety and agricultural health standards: Challenges and opportunities for developing country exports", *Report No.* 31207.

———. (2007). "India: Taking agriculture to the market", *Report No* 359530-IN.

Annex: Insights in High-value Commodity Chains[1]

Annex I

Fruits:

The Case of Apples

I.1. Methodology

Insights into the functioning of the value chain of apples were obtained during discussions with different actors during our visit in the state of Uttarakhand. Interviews were held with producers, representatives of farmer groups (the Fruit Growers' Association), consolidators, representatives of Mother Dairy, managers and members of local NGOs in the district of Nainital and Uttarakashi and with processors, wholesalers, retailers, commission agents and representatives of the APMC in the wholesale market of Haldwani and Dehradun. We report most of the findings of these interviews in this Annex, starting from the producer level to primary marketing options, organisation of the wholesale markets, and retailing. We complement these findings with a literature review as well as with primary analysis of Faostat and Agmarknet data.

I.2. Production

I.2.1. The Indian Situation

Indian apple production was about 1.35 million tonnes in 2005. The area of apples planted is about 250,000 hectares, the second largest in the world. The average yield is evaluated at 5.5 tonnes per hectare, the lowest of the major world producers (Deodhar *et al.*, 2006). Although harvested areas have expanded since 1990, declining average yields have slowed annual production growth to about 1.5 per cent (Deodhar *et al.*, 2006 and Figure I.1).

Figure I.1 **Trends in Apple Production in India 1990-2005**

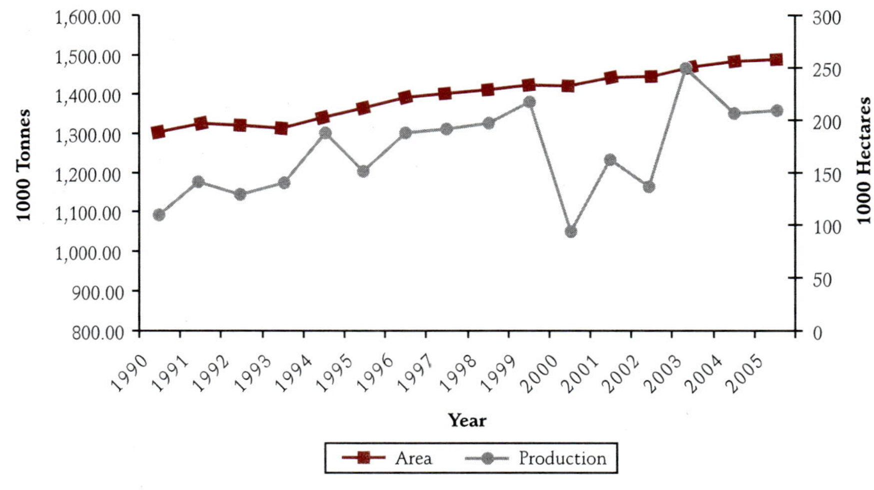

Source: Faostat.

1. Given that the analysis in the value chain studies is largely based on insights of key informants or focus groups, we would like to caution readers to the bias that this might cause for the analysis. Given that few updated agricultural surveys at the state level are available, a representative statistical analysis at the state level was not possible within the time period of the study. While we were careful in choosing our key informants and tried to triangulate the information given (based on available statistics and on the analysis of the Agmarknet data), this was however not possible in all cases.

As apples can only be grown in temperate climates, the majority of the Indian production is situated within three states (Jammu and Kashmir, Himachal Pradesh and Uttarakhand), in regions ranging in altitudes between 1,000 and 4,000 metres. While statistics on apple production and area estimation in India might be questionable (Deodhar *et al.*, 2006), it is assumed that the state of Jammu and Kashmir is the biggest producer in India, counting for about 65 per cent of the production in a regular year. The second is Himachal Pradesh (20 per cent of the production) followed by Uttarakhand (10 per cent of the production). Other states count for about 5 per cent of total production (Deodhar *et al.*, 2006).

Different types of apple varieties are grown in India. Most of them are variants of the Red Delicious and the Royal Delicious varieties. Although more than 700 accessions of apples introduced from abroad have been tried and tested during the last 50 years, Deodhar *et al.* (2006) estimate that the Delicious group of cultivars still account for about 83 per cent of the production in Himachal Pradesh and more than 45 per cent in Jammu and Kashmir.

Apple production in India is highly seasonal. The bulk of the local harvest happens between September and October. There is a slight difference in the timing of harvests between the different states because of differences in altitude and latitude. The harvest peak months are in July-September in the state of Uttarakhand, in August-September in Himachal Pradesh and in August-November in the state of Jammu and Kashmir. Cold storage is prevalent in the state of Jammu and Kashmir as they are the last producer and they might thus reap the benefits, at least cost, of the increasing prices after their harvest.

I.2.2. The Situation in Uttarakhand

Statistics on apple production and areas planted vary widely in Uttarakhand. The Agricultural Statistics Division (GoUA, 2007) estimated the area planted in 2001 at 3,174 hectares, based on the results of the census of 2001.[2] Lal (2003) evaluates during 1999-2000 the area at 55,960 hectares and the production at 60,495 tonnes. The Directorate of Horticulture relies on compilations of data at the district level by their district level horticulture officers. It estimates the areas planted in 2005-06 at 28,800 hectares and the production at 112,320 tonnes. While this Directorate believes that about 50,000 hectares are planted in total in the state, it estimates that only about 28,000 hectares are productive. Mariental (2006) reports that apple production is about 180,000 tonnes under the most 'congenial weather conditions'. Obviously, this large variation in statistics and the uncertainty surrounding production should be of large concern to policy-makers and (potential) stakeholders as it makes any planning or investments hard to do.

The main varieties that are found in the state of Uttarakhand are: 1/ in the early season (harvest in July or late June): Early Shanburry, Fanny and Benoni. These varieties count for 30 per cent of the area under cultivation; 2/ in mid season (harvest in August): Red Delicious, Starking Delicious, Spritzenburg, Golden Delicious. These varieties occupy 50 per cent of the area under cultivation; 3/ in the late season (harvest in September): Rymer and Buckingham, counting for 20 per cent of the area (Lal, 2003).

The major advantages of apple production in the state of Uttarakhand compared to the other two producing states are twofold (Lal, 2003): First, apples from Uttarakhand reach the market about a month sooner than the apples from other states and secondly, the apple orchards of the state are almost free from scab disease.

The major reported production constraints are several. The ecological constraints in the state of Uttarakhand include (Lal, 2003): 1/ water scarcity (fruit development is hampered during the summer season when there is generally a scarcity of soil and atmospheric moisture); 2/ steep terrain (hampering the adoption of scientific orchard management); 3/ changes in temperature (sudden changes affect the flowering period); 4/ hailstorms (the incidences are frequent during the month of April causing early fruit drop or damaged fruits).

---

2. In the plain areas of the state, land record forms and registers are prepared on a complete basis while in the hills, a sample of 20 per cent of villages at random is taken in such a way that during a period of 5 years all the villages in the hills are covered (GoUA, 2007).

The managerial constraints include (Lal, 2003): 1/ obsolete varieties (which fetch a low market price); 2/ lack of pollinisers (both pollinisers and pollinators are not provided up to the desired level; the tendency of poor bee keeping is a also a major constraint); 3/ poor soil status and management (it is customary to plant apples on poor soils while rich soils are used for growing cereal crops); 4/ incorrect training and pruning (plants need correct training in early stages and scientific pruning later on); 5/ distant plantations (planting is done for the conventional planting system on vigorous rootstock; high density plantations have not yet been adopted).

Losses at the farm level can be high and focus groups of farmers estimated that on average 15 per cent to 20 per cent of the normal production could be destroyed. These losses could be due to a variety of reasons, linked to storage problems, hail storms, and losses due to insects and birds. Hailstorms would happen every year in pockets of the state and some farmers would be able to protect themselves against the damages caused by these through the use of protective nets. However, usually only the richest, largest farmers would (be able to) take these preventive measures.

While no exact data are available, most stakeholders feel that apple producers are relatively richer farmers. This is largely due to the amount of land that is required for planting orchards as well as the due to the long waiting periods before investments bear fruit. Apple orchards are often considered a status symbol in the area. A focus group in Naugoan estimated that 60 per cent of the apple farmers would cultivate between 2 and 5 hectares of apples while 40 per cent of farmers would cultivate 1 hectare. Only rich farmers can allow for the wait of investments in apple orchards. It is estimated that it would take about 5 years before production can begin for regular varieties. For the better valued Delicious variety, it would take even between 8 and 10 years. However, it seems that this waiting period could be reduced significantly. There are now modern technologies available where production could even start after one year.

The productivity per tree changes dramatically over the years as trees mature. For example, production per tree would increase from 250-300 apples after 5 years to 500-800 apples after ten years. Apple production then goes down when apple orchards become too old (an important reason for low productivity in parts of Uttarakhand). The number of trees per hectare planted in the State is estimated to be on average between 200 and 500. However, new (Dutch) technologies are now available that should allow an increase of the number of trees per hectare to 800. If properly done, this would allow for a significant increase in land productivity.

Farmers rely heavily on pesticides in apple production. Even though the price of pesticide has gone up over time, the use of pesticides has increased significantly. Focus groups stated that farmers used to spray between 2 and 4 times a season but that this has now sometimes gone up to even about 20 times a season. Some farmers do realise the benefits of organic farming and are involved in organic production; these farmers report to receive a price premium of 10 to 15 per cent above the other apples for doing so.

Changing weather patterns (drop in the quantity of snow; less rainy days) might further take their toll on apple production in the state. Given the required number of chilling days, apple production might over time only be possible in the upper reaches (7000+ ft) as lower reaches might not be able to sustain the ecological demands of the crop. Other crops (such as pear or walnut) might be better able to deal with weather vagaries, especially so in the lower reaches. On the other hand, new varieties are becoming available that do not require this long chilling period which would give a potential to apple production in lower reaches (3000-4000 ft). It seems that farmers in the lower reaches are eager to adopt such varieties.

I.3. Marketing Channels

Modern retail buys few apples from the state of Uttarakhand. It seems only Mother Dairy does so. However, the quantities are limited. Mother Dairy procures apples in the Nainital area for two uses: processing and table varieties. The prices differ widely between the two: This year, it paid Rs. 20-22/kg for the table variety and about Rs. 3.25/kg for the processing variety. The prices offered in the *mandi* for the processing varieties might even be lower than what Mother Dairy offers. The apples procured by Mother Dairy for the table varieties are better quality, i.e., the A-grade, implying that apples need to be at least 45 mm in size and that they are red and shiny. The only requirement for the processing varieties is that they are not rotten and are free of insects. Any variety and any size will thus be purchased. Farmer groups stated that the prices that Mother Dairy offered were not significantly higher this year than the prices offered on the local *mandi*.

Processing of apples in the state is limited. There exists about 48 government and community training centres where the people bring their produce for local processing. This is mostly done for local consumption and the processed apples are not for sale in the market. Other than Mother Dairy, there are also some NGOs that are active in the processing of apples for juice and jams. However, the quantities that are processed, are limited.

There is further a government initiative to procure apples at a Minimum Support Price (as is the case in the state of Himachal Pradesh or in the state of Jammu and Kashmir) as stated in the 11th Plan. The procurement was done in 2007 at a price of Rs. 4/Kg by KMVN and the GMVN. These apples are then sold to local processing centres. However, it is estimated that the quantity of apples that is procured that way is limited. The Department of Horticulture estimates its importance at 5 to 7 per cent of total production. None of the farmer groups that we talked to would consider this a potential outlet for their produce, indicating its current limited importance.

The quality of the apple leads to significantly different prices in the market (even in the traditional channels). Apples are classified in a system of A, B and C grades (based on colour and size). The price differences between the grades are large. It is estimated that, last year, A-grades were valued at Rs. 50 to 60/Kg, B-grades at Rs. 10 to 20/Kg and C-grades at Rs. 4 to 5/Kg. While no exact data are available, there are seemingly large quality differences between the different States. It was estimated by the farmers that the production in Uttarakhand could be divided between 40 per cent A-grade, 30 per cent B-grade and 30 per cent C-grade. The quality produced in other states would be significantly higher. For example, it is estimated that in the state of Himachal Pradesh, 60 per cent of the apple would be A-grade, 20 per cent B-grade and 20 per cent C-grade. The C-grade is used almost exclusively towards processing. About 30 per cent would go to the processing industries in Himachal Pradesh (20 per cent of the total production) or to Darbur (10 per cent of the total production for jams). However, most of the C-quality (70 per cent) would be used for small-scale local processing industries.

Many different packing materials are used in domestic apple trade. The most common are thin wooden crates of 10kgs (the most used in Uttarakhand) or 25kgs with straw packing. The cost of these wooden crates is about Rs. 25 for a small box and Rs. 35 for the larger box. While the government has started to distribute carton boxes (CFB boxes), at subsidised prices where farmers could buy them for around Rs. 10 per box, there seems however to be regional variability in the adoption of these boxes. While in the western part of the state, carton boxes have been widely adopted, farmer groups in the central part (Nainital) stated that there was only limited supply of these boxes and that they were not suited for transport in the hills. Use of high-quality boxes with sufficient rigidity to protect the fruit is overall limited in the state. The high costs of packing material, borne by the farmer, leads to significant lower prices for the farmer.

While there are only a limited number of commission agents in the state who procure apples, there are also a large number of commission agents from outside the state who deal with apple trade within Uttarakhand. It was estimated by focus groups that 90 per cent of the apple farmers would accept input advances from these commission agents. Most of the farmers are able to obtain information about prices as they use mobile phones to obtain information and it seems that they are able to switch to these commission agents that offer the best prices, despite the input advances (some commission agents reported that a large number of farmers would not repay the input advances). Selling of apples before harvest is sometimes done (it is not clear how common this practice is). This leads some farmers to take less care of the production process as they have little incentive to do so.

India is also involved in international apple trade. Figure I.2 shows the evolution of apple trade in the last fifteen years. While imports and exports are relatively small compared to total domestic production, their importance is quickly rising. Apple imports have especially taken off since the removal of quantitative import restrictions in 1999. The biggest sellers/source countries are the US, New Zealand/Australia and the People's Republic of China. The imported apples are a far superior quality than the locally produced apple. US apples imported by India are primarily Washington Red Delicious. Apples from Australia and New Zealand are mostly

Gala and Red Delicious varieties. Chinese apples are primarily of the Fuji variety (Deodhar *et al.*, 2006).

Indian apple imports follow a clear seasonal pattern with few imports arriving during the local peak harvest months and rising imports during the lean period of December-July (Deodhar *et al.*, 2006). The bulk of the imports arrive during the months of April to June. It seems that importers choose to import little during the local harvest periods because of the large drop in local prices. While imported apples command a premium over locally produced apples due to their higher quality, the high import prices present a less attractive opportunity for the importers to compete during that period (Deodhar *et al.*, 2006).

Figure I.2 **Imports and Exports of Apple in India**

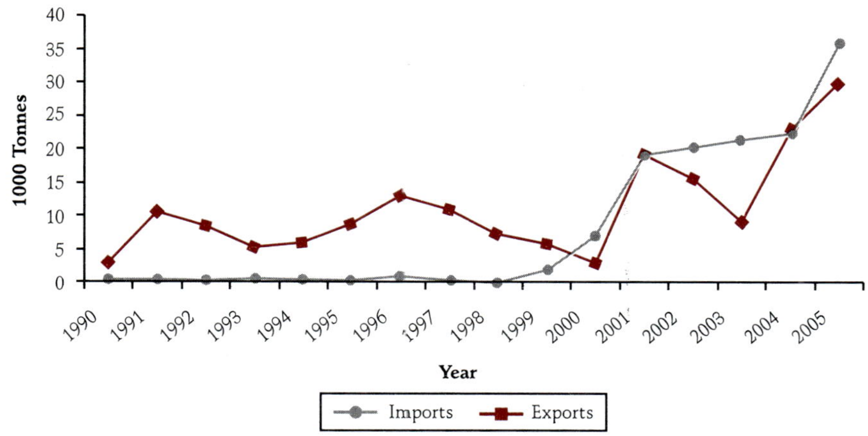

Source: Faostat.

Import regulations ensure that imported apples are expensive in India (Deodhar *et al.*, 2006). First, India's import tariffs on apples (50 per cent) are the highest among the major fruit that it produces domestically. Second, India has a number of plant health and food safety regulations that apply to apple imports. Importers face transaction risks due to the uncertainty regarding the interpretation and enforcement of the regulations. These regulations concern mostly plant quarantine, pesticide residues and apple waxing.

While imports are on the rise, exports have also increased significantly since 2000 (Figure I.2). Bangladesh has been the principal importer from India accounting for more than 75 per cent of the total exports (Mariental, 2006). India has also been exporting apple juice and apple juice concentrate to countries like the US and the UK. There has been an almost nine-fold increase in the quantity of juice exported from India despite the strong competition from the People's Republic of China (Mariental, 2006). However, the exported quantities are still rather small, especially compared to local production.

I.4. Wholesale, Retail and Demand

The majority of the apple trade in Uttarakhand takes place on the Haldwani market (39 per cent of the arrivals) and the Dehradun market (30 per cent of the arrivals). Together, they accounted for 70 per cent of all arrivals in the *mandis* during the period April 2006–March 2007. Figure I.3 shows the monthly arrival of apples on all the *mandis* in the state of Uttarakhand in the period April 2006–March 2007. It illustrates the high seasonality of apple trade. The majority of arrivals is in the month of July after which there

Figure I.3 **Apple Arrivals in *Mandis* in Uttarakhand (April 2006-March 2007)**

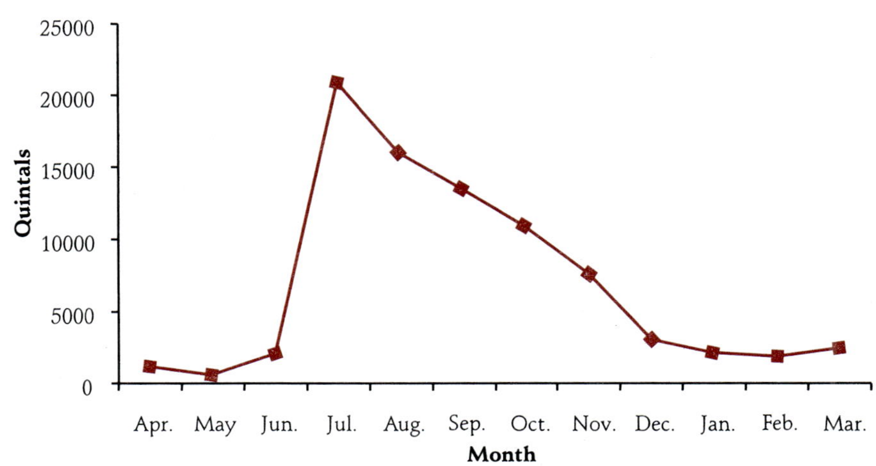

Source: Own calculations based on Agmarknet data.

is a gradual drop until the month of December. Note that arrivals in the *mandi* do not only concern apples that are produced in the state of Uttrakhand. It seems that the largest part of the arrivals in the months of October and November is from production out of state, which might then be consumed locally.

At the *mandi*, apples are traded by commission agents. Commission rates charged to the farmer are different between different states and *mandis*. Farmer focus groups indicated that commission agents from Uttarakhand would charge the highest commission to farmers: 10 per cent. Second would be commission agents from Delhi: 8 per cent, followed by commission agents from Chandigarh who would charge 6 per cent.

Cold chains are widely used in the state of Himachal Pradesh. This allows for longer storage of apples. That state is also planning to set up a cold storage facility of 8,000 tonne along the Haryana-Delhi border (to be ready before the next apple season) as to allow apples from Himachal Pradesh to enter the Delhi market at the right time. In Uttarakhand, on the other hand, there are currently no cold chains. This seems driven by different reasons: the quantities produced are significantly lower; the produced quality makes storage unprofitable; and the scattered farmers would make an appropriate location of cold storage problematic.[3]

Figure I.4 **Apple Consumption in India 1990-2005**

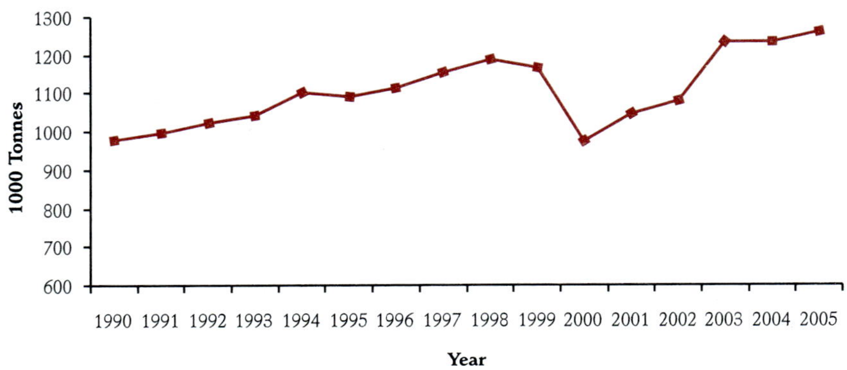

Source: Faostat.

Figure I.5 **Wholesale Apple Price Indices 1993-2006**

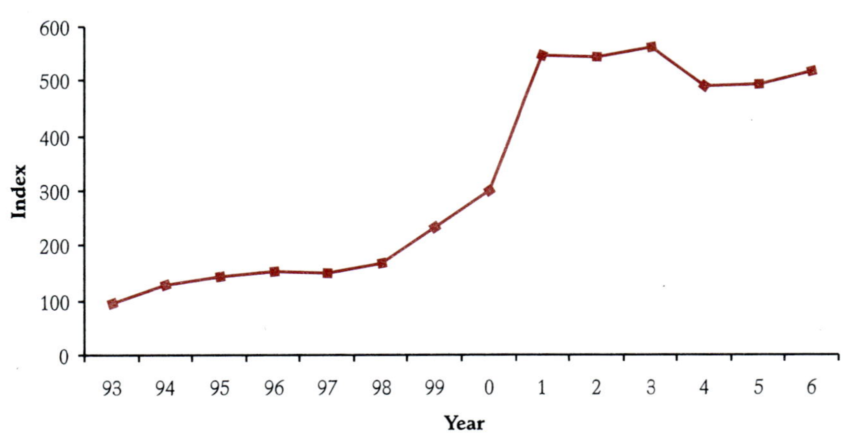

Source: Office of Economic Advisor, Ministry of Commerce.

Figure I.4 shows the evolution of the apple consumption in India between 1990 and 2005. Except for the years 2000-2002, the trend is slowly upward. If we compare the overall trend with the growth in population, per capita apple consumption did not increase over time. Because apples are a relatively high-priced fruit in India, consumption is largely confined to the higher-income segments of the population. This is confirmed by detailed demand analysis. Devadoss and Wahl (2004) report an income elasticity of 1.05 for domestic apples in India, implying that a 1 per cent growth in income would lead to a 1.05 per cent increase in consumption. They further estimate a high own price elasticity of -0.53, indicating that a 1 per cent increase in prices would lead to a 0.53 per cent reduction in consumption. Apple consumption is thus highly dependent on prices and income.

While consumption has remained stable, apples prices seem to have increased dramatically in the last decade (Figure I.5). Using the wholesale apple price index published by the Ministry of Commerce, apple prices in 2006 were 5 times higher than in 1993. However, the increase in apple prices happened especially in

3. It was estimated by focus groups that Uttarakhand apples can get a rate as high as Rs. 25/kg but could drop quickly as low as Rs. 8/kg when the harvest of Himachal Pradesh kicks in, as the quality of the apple from Uttarakhand is significantly lower than the one of Himachal Pradesh and Jammu and Kashmir.

the short period between 1998 and 2001. Apple prices have been stable between 2001 and 2006 and have even shown a slight dip in the years 2004-2006 compared to three years earlier.

Apple prices are further characterised by significant seasonality, linked to production periods. This is shown in the example of the *mandi* of Dehradun and Delhi below (Figure I.6). Prices are lowest during the major harvest periods in the country (September–October) and prices are 25 per cent to 30 per cent higher during the rest of the year. Given that the majority of the harvest of apples in Uttarakhand falls in the months of July-August, the state is able to benefit from the slightly higher prices that are offered during that period. The graph also shows that cold storage of apples from Uttarakhand would not be competitive: First, the purchase prices are higher than in other producing states at the time of harvest; Second, apples would have to be kept longer in storage than in other states as prices first decline before they increase. The opportunity cost of capital would then make such a venture unattractive for potential investors.

Figure I.6 **Apple Prices on Wholesale Markets in Uttarakhand and New Delhi (April 2006-March 2007)**

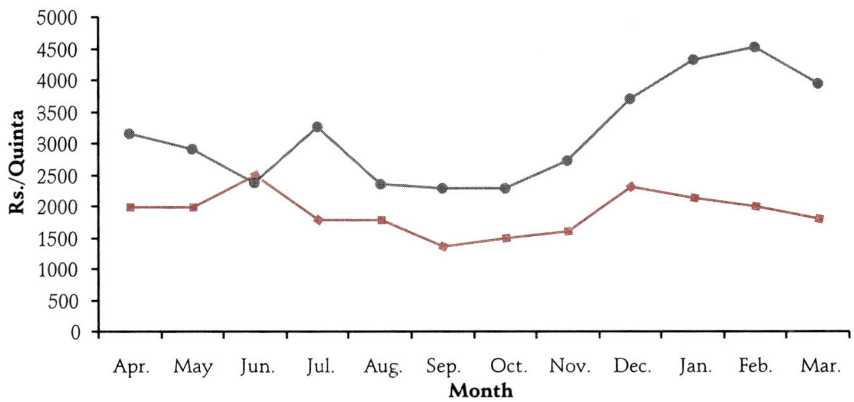

Source: Own calculations from Agmarknet data.

Figure I.7 further illustrates this high seasonality in production and prices in one of the most important terminal markets for apples in India, New Delhi. The majority of the apples are traded in the period from August to November. This period also coincides with the lowest price of the year. The graph shows that apple prices are almost twice as high in the lean period compared to the harvest period.

Figure I.7 **Apples Arrivals and Prices on Azadpur-New Delhi (July 2006-June 2007)**

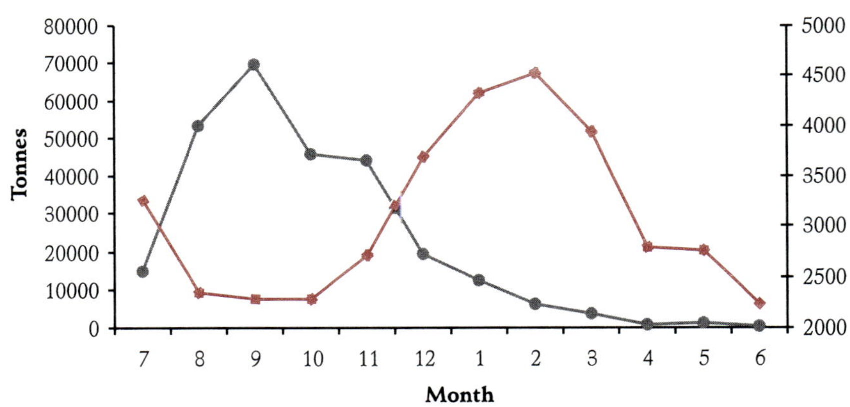

Source: Own calculations from Agmarknet data.

Agmarknet collected price data of apples between 2003 and 2006 in the major *mandis* in Delhi. It also collected information on quality and size. Using these data, we look at the rewards that exist in the wholesale *mandis* for variety and size. A regression was therefore done of the log of the modal price (noted on the market by Agmarknet) on the size of the apple, the reported variety, and monthly, yearly and market dummies. Using these results, we are able to have quantitative ideas on the premiums that traders are willing to offer for variety and size.

We find that rewards to the size of the apples are quite large. Medium sized apples fetch a price that is 27 per cent higher than small apples while larger sized apples are sold at a 49 per cent premium. Lal (2003) finds that the majority of the apples that are marketed in Uttarakhand are smaller-sized (60 per cent of the supply) and it thus seems that farmers in the state miss out on the premiums that are offered for size. It seems special efforts are needed to enhance fruit size by adopting better orchard management practices, fruit tinning and the applications of nutrients (Lal, 2003).

Figure I.8 Price Premium for Different Apples Variety Compared to the Delicious Variety

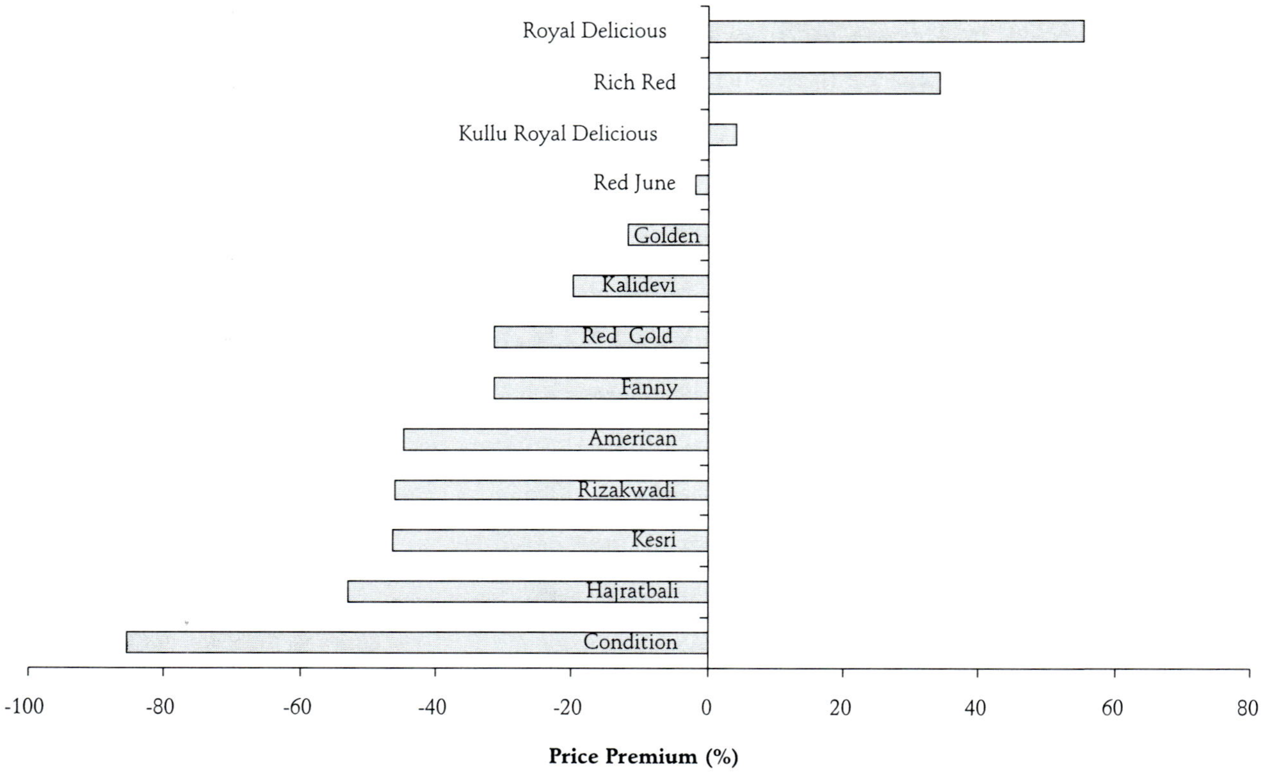

Source: Own calculations based on Agmarknet data 2003-2007.

We also find large differences in prices between varieties (Figure I.8). Only the Royal Delicious and the Rich Red varieties fetch prices that are significantly higher than the Delicious variety (the most common variety). All other varieties are valued significantly less than the Delicious varieties. The prices that are in the Agmarknet data do only include table varieties as processing varieties are not traded in Delhi *mandis*. The prices for these varieties are the lowest of all (such as the Rymer variety that consumers do not prefer to eat as table variety because of its sour taste). The different values for different varieties would thus give an incentive to switch to the more valuable varieties. While the area under Delicious is about 93 per cent in the state of Himachal Pradesh, about 50 per cent in the state of Jammu and Kashmir, it is only 30 per cent in the state of Uttarakhand (Rathore, 2003). Most of the apple farmers that we talked to were aware of these price differences in the market and they were planning to take them into account in future planting. However, some of the varieties that fetch high prices might be hard to cultivate in Uttarakhand. For example, the Delicious varieties that are on the market need a chilling period of 45 days for optimal production which the physical environment of Uttarakhand can seemingly not provide, unless in some specific areas.

During our field visits, we conducted a large number of interviews with traders as to better understand marketing costs and see how they contribute to price formation. The marketing costs for a low quality—and thus lower priced—apple is significantly different than a high quality apple. Using an average of results coming out of our interviews, we find that the marketing margin makes up Rs. 17.7/kg or 50 per cent of the retail price in the cities of Uttarakhand in the case of the high quality apple. These numbers illustrate the cost of marketing in Uttarakhand. This share would probably increase for the produce that is marketed in other states or cities. A big part of the marketing margin goes to the commission rates and the taxes that charged by the commission agents. This amounts to 21 per cent of the total marketing margin in high quality apples. The retail margin contributes Rs. 7/kg or 41 per cent to the total marketing margin between producer and consumer. Transport costs are relatively less important accounting for 6 per cent of the marketing costs.

Wastage by retailers was estimated to be around 10 per cent. Interestingly, the majority (about three quarters) of the wastage at the retail level was reported immediately after the purchase of the product. There is a common practice to place under-sized and mis-shapen fruits on the bottom layers of packing cases and good fruits on top (Lal, 2003). It is not unlikely that farmers also put already damaged fruit in there at the time of packing. This is likely the case as the interviews were conducted with retailers that were close to production sites and transportation would presumably not have taken such a heavy toll. Retailers would take on average four days to sell the produce and would report a loss of the other quarter (2.5 per cent of the total) because of sales problems. The wastage numbers on domestic apples is in strong contrast with imported apples where no wastage was reported by the retailers that we talked to.

In contrast to domestic apples, imported apples maintain their quality largely because of a superior cold chain (Deodhar *et al.*, 2006): Imported apples remain in their refrigerated container until they reach their major urban destination and are then held in cold storage until they are sold to a wholesaler. Imported apples also command a high price premium because of the uniformity in size, colour and shape as well as the low levels of latent damage due to higher quality packing (Deodhar *et al.*, 2006). Domestic apples are typically transported, handled and stored in poor quality packaging with higher levels of damage to the fruit.

However, it seems that due to the high prices that are reported for imported apples and due to the increasing prices for domestic apples, domestic private firms are beginning to stimulate quality improvements. For example, the Container Corporation of India (Concor) was planning to procure 12,000 tonnes of apples from Himachal Pradesh in 2007 through its cold chain subsidiary Fresh and Healthy (*Hindu Business Line*, July 29th, 2007). It would supply these apples after the main harvest periods to retailers such as Reliance Fresh, Big Apple and Mother Dairy.

I.5. Conclusions

Apples are the most important fruit of the state of Uttarakhand. While apples benefit from some advantages in the state (such as the early production period compared to the other states as well as being free of the scab disease), the value chain faces significant challenges. Most of the apples that are produced in Uttarakhand are of low quality. However, they do however fetch relatively high prices as they enter the market when only imported apples are available. While the production of apples in Uttarakhand is agro-ecologically challenging especially given the changing climatic situation (as apples need the right complicated mix of light, temperature and altitude), some managerial practices could be improved that would help fetch higher prices for its producers.

A focus on better varieties in the state might have a large pay-off. Some varieties are now available that could be grown in the lower reaches of the state (as these varieties do not require the long chilling periods). Some of the varieties that are currently grown are also not very highly valued in the market. However, it seems that there is a trend to improve varieties as to better capture the demands in the market. Short-term interventions that may benefit the farmers would be to improve the regulatory and competitive environment and reduce the packaging costs by promoting better vertical integration.

While the bad situation of the apple sector in Uttarakhand is often compared to the better organised states of Himachal Pradesh and Jammu and Kashmir, it seems important to consider that these states are significantly more dependent on apples while in Uttarakhand, it is only one crop among many that is also often grown by relatively better-off farmers. This means that Uttarakhand would not be able to benefit from economies of scale emanating from research, extension, cold storage investments, and other value chain interventions that these other states are heavily involved in.

Annex II

Vegetables:
The Case of Potatoes

II.1. Methodology

Interviews were held with producers, representatives of farmer groups, consolidators, representatives of Mother Dairy, representatives and members of local NGOs in the district of Nainital and Almora (INHERE, Chirag) and with wholesalers, retailers, commission agents and representatives of the APMC in the wholesale market of Haldwani. We report most of the findings of these interviews in this annex, starting from the producer level to primary marketing options, organisation of the wholesale markets, and retailing. We complement the results of the interviews with a literature review and a primary analysis of Faostat and Agmarknet data.

II.2. Production

II.2.1. The Indian Situation

Potato is an important crop in India. India is ranked third in production in the world after the People's Republic of China and the Russian federation. Potato is cultivated in India under diversified agro-climatic conditions. The following three crops are raised every year: (a) *Kharif* crop: From July-August to October-November (12 per cent of total production); (b) Winter crop: From October-November to February-March (80 per cent of total production); (c) Summer crop: from March-April to July-August (8 per cent of total Indian production).

Climatic variations across India determine not only the regional distribution of the crop, but also to a large extent the seasonal patterns. The optimal temperature range for potato is from 15°C to 25°C. However, night temperature is critically important to tuber development and should not exceed 20°C, a factor which limits the season in many areas in the south. Frost can limit the season in the higher altitudes of the north (CIP, 2006). A major challenge in India is potato storage as the potato production is followed by hot summer months which makes refrigeration necessary for storage.

Cold storage takes place on a large scale and it is estimated that there are currently over 3,400 cold storage facilities in India that can store over ten million tonnes (or about 40 per cent of the production) (CIP, 2006). CIP (2006) further estimates that approximately 60 per cent of potatoes in cold storage are ware potatoes, intended for consumption while the other 40 per cent is used for seed. Storage of ware and seed potatoes usually happens at different temperatures. Storage for seed should best be done at lower temperatures (2°C-4°C) as to avoid premature sprout growth. However, cold temperatures also induce high levels of sugar accumulation, rendering potatoes less suited for consumption.

Over the past few decades, potato has become the fastest growing staple crop in India (CIP, 2006). Per capita production has increased by 372 per cent between 1961-1963 and 2001-2003. The

Figure II.1 Trends in Potato Production in India 1990-2005

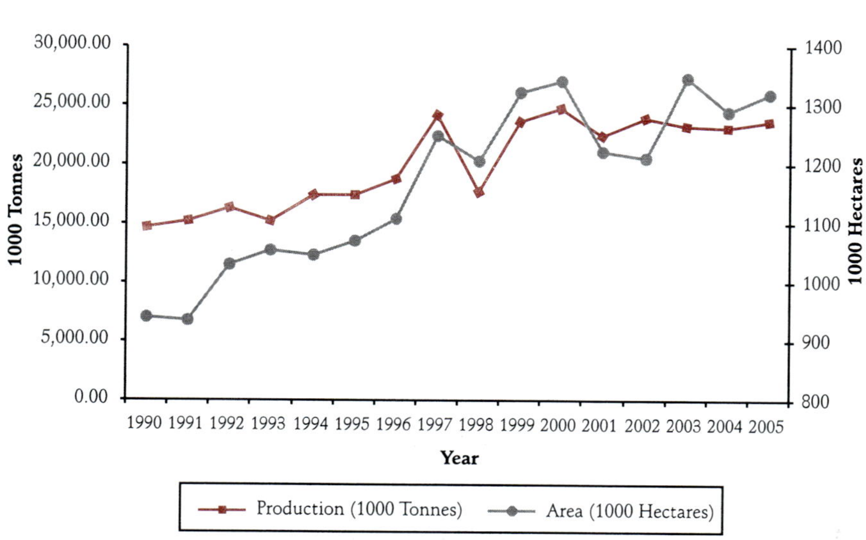

Source: Faostat

increase is driven in part by a shift in food habits, as demand in potatoes has increased due to relatively higher incomes. This rising trend has continued in recent years (Figure II.1). Between 1990 and 2005, potato production has increased by 60 per cent and the area allocated to potatoes has increased by 40 per cent. The lower expansion of area compared to production indicates a slight improvement in productivity, increasing from 16 tonnes/ha in 1990 to about 18 tonnes/ha in 2005. CIP (2006) finds that especially urban Indians with higher incomes are providing much of the growing demand for potatoes. The potato in India is not primarily a rural staple but is a cash crop which provides significant income to rural farmers.

India only exports a limited amount of potatoes. It exports around 300,000 tonnes of fresh, seed potatoes and processed potatoes (MCX, 2006). Nepal, Sri Lanka, the United Arab Emirates and Mauritius are the major destinations for fresh potatoes. The major destinations for Indian seed potatoes are Nepal, Sri Lanka and the United Arab Emirates.

II.2.2. The Situation in Uttarakhand

Potato is one of the major vegetables being produced in the state. Production was estimated in 2005-06 at 450,637 metric tonnes produced over an area of 22,157 hectares, i.e., an average yield of 20 tonnes/ha. Potato is an important cash crops for smallholders as it is often grown by them. For a large number of farmers, potato is also still a major subsistence crop grown for own consumption.

Farmers usually use tuber seeds from their own production or buy seeds on the *mandi*. Seed inputs make up a significant cost of total input costs as 1 kg of seeds produces only 7 to 8 kgs of potatoes. While the majority of the farmers rely on own seeds, they purchase new ones once every three years (through shops at the mandi). Fertilisers are regularly used by farmers for potato production. Insecticides (Furaden) or fungicides (often Dithane M45 produced by Dow Agro Sciences) are commonly used, especially in periods of heavy rains. The total monetary cost of modern input use was evaluated by farmer focus groups at around Rs. 0.50 /Kg. Fields where potatoes are grown are usually cultivated twice a year with *madua*, cabbages or other crops being grown during the other season. A common complaint by hill farmers is that potato yields are reducing over time. This is linked to climatic reasons due to low rainfalls and higher temperatures. Both are seemingly bad for their potato production. A major production problem for potatoes, as well as for other crops, in the state of Uttarakhand is wildlife (boars) that eat potatoes before they can be harvested.

Only few post-harvest activities are reported by potato farmers. These activities are drying of the potatoes in the sun and sometimes grading of the potatoes before taking them to the market. The drying in the sun will help in the maturation of the potato skin which will reduce bruising and damage during transportation and storage. The fact that potato tubers consist of about 80 per cent water makes them extra vulnerable to storage losses (Khatana, 2003). Cold storages are of comparatively recent origin (in relation to potato cultivation). Storage of potatoes at lower temperatures in cold storage converts starch to sugar. Such potatoes are not fit for processing and they are also less preferred by the consumer. Different indigenous storage methods exist in the country and these methods allow farmers to keep potatoes 'fresh' for a period of up to 2 or 3 months. Uttarakhand grows potatoes in different agro-ecologies and farmers use a variety of methods for temporary storage of up to three months. The methods are influenced by a number of factors such as climatic factors and storage losses are related to a large number of factors. However, farmers often inspect the tubers and get rid of tubers which look like they may rot. Such potatoes are consumed in the house and effective losses are thus minimal (Khatana, 2003).

II.3. Marketing Channels

Modern retailers who buy up potatoes in the state of Uttarakhand are limited. Mother Dairy (MD) purchases potatoes through different outlets in Uttarakhand. As is customary practice for modern retailers, it has strict requirements on the quantity and quality that it purchases. Mother Dairy only buys those potatoes that are of high quality. This means they have to be of medium-size (between 45 and 50 mms) and free of insects and spots. There is little large-scale potato processing in the state of Uttarakhand. There are numerous small-scale units though that process for localised consumption. The only large company that is active in the state is ITC.

It has a potato processing unit in Haridwar where it produces the 'Bingo' potato chips. However, it procures most of its potatoes for this plant from the major producing areas around Agra and only occasionally—in the case of shortfalls—purchases produce from the local *mandi* in Haridwar.

A large number of hill farmers sell potatoes immediately after harvest (and keep a little bit for their own consumption) and they then purchase potatoes during the rest of the year. This is typical behaviour for farmers that have rather easy access to markets. This behaviour was noted by farmers as well as APMC authorities. For example, APMC authorities in the Haldwani *mandi* noted that hill potatoes were marketed during the period of May 15th until October 15th and during that period, hill potatoes are being exported from the Haldwani *mandi* to UP markets or to Delhi. In the period between Ocober 15th and May 15th, the hill areas would be deficient in potatoes and potatoes would be imported from the plains for the consumption in the hills.

We tried to arrive at the marketing costs of different potato farmers that sold potatoes to the *mandi* in Haldwani. The farmers that we interviewed included farmers from producing areas around Simayil (Nainital) and other farmers from around Bageshwar. All farmers sell potatoes in bags at the *mandi*. They are charged 5 per cent commission rates by the commission agents. Additionally, they are charged 1 per cent sampling costs (farmers are unclear on the usage of this but think that part of this is given to temples and part goes to the APMC authorities) as well as communication and labour costs.

The farmers in Simayal estimated that about 95 per cent of their produce was sold at the regular *mandi* compared to 5 per cent to Mother Dairy. To transport their produce, they would pay Rs. 38 for a bag of 100 Kgs. It would take these farmers 4 hours to transport their produce to the Haldwani *mandi*. On a usual trip, they would leave very early in the morning and come back at night the same day. The time needed for the transaction of their potatoes varies a lot. Sometimes their lots can be sold in half an hour. Other times, it takes 5 to 6 hours and sometimes farmers have to stay an extra day before all is sold. Most of the commission agents that these farmers deal with would pay the farmers immediately.

The farmers from Bageshwar came on a truck with 4 farmers. The distance between their home and the market was more than 300 kms and the farmers left the day before to come to the market. The transport costs that they had to pay to the transporter were between Re. 0.50/Kg (Rs. 25 for 50 Kg) and Re. 0.375/Kg (Rs. 750 for 2 tonnes). The farmers also had to pay for local transport costs to transport potatoes from their village to the road. These costs amount to Re. 0.625 /Kg (Rs. 25 per 40 Kg). The time between transport/marketing and harvesting would be 2 months. One of the farmers that we interviewed stated that about 80 Kgs out of 2 tonnes were damaged during the trip (about 4 per cent). Most of the bags that he used were his own bags which he stated could be used for over 4 years. The farmer would always go to the same commission agent. He could have access to advances with the commission agent but he did not ask them.

Figure II.2 Potato Arrivals in *Mandis* in Uttarakhand (April 2006-March 2007)

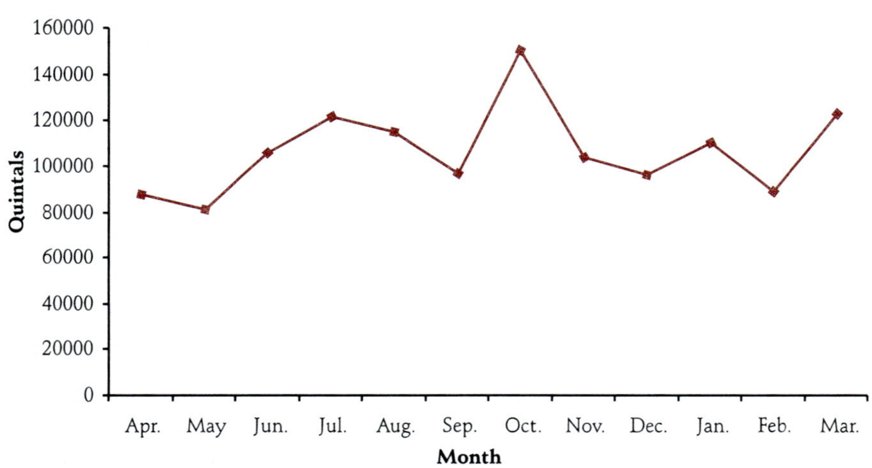

Source: Own calculations based on Agmarknet data.

II.4. Wholesale, Retail and Demand

To understand the functioning of the potato wholesale market, we rely on Agmarknet data on arrivals and price formation in all the *mandis* of the state. In Figure II.2, we plot the arrivals of potatoes based on the Agmarknet data in the period between April 2006 and March 2007. There is surprisingly little seasonal movement in the quantities that are sold as a whole in the *mandis*. The highest point is the month of October when over 140,000 quintals are sold. This compares to a low of 80,000 quintals in the month of May. Given the strong seasonality in production patterns in different districts of the state, it thus seems that we have strong changes in flows between different months of the year that allow for equilibration of the quantities over the year.

Figure II.3 **Seasonal Patterns in Potato Arrivals in Hills and Plains in Uttarakhand (April 2006-March 2007)**

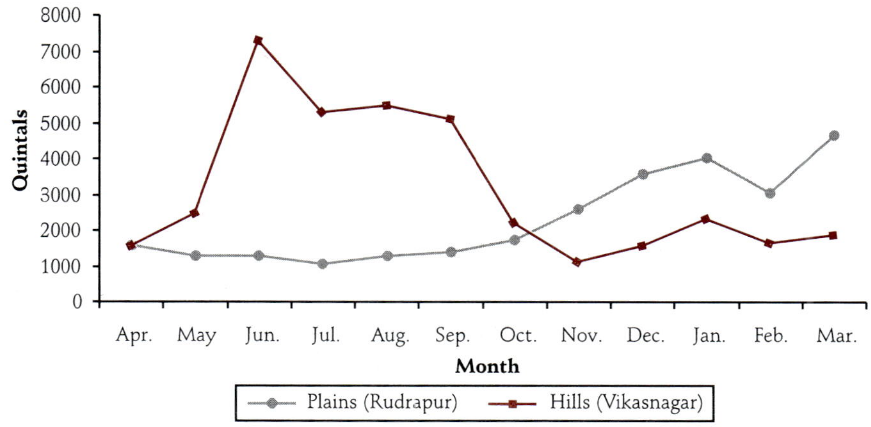

Source: Own calculations based on Agmarknet data.

To illustrate the differential production and sales patterns between hills and plains, we contrast the sales patterns of two *mandis* of similar magnitude in the plains (Rudrapur) and in the hills (Vikasnagar) (Figure II.3). The total sales during the year April 2006–March 2007 amount to 27,540 quintals in the Rurdrapur *mandi* and 38,251 quintals in the Vikasnagar *mandi*. Seasonal patterns are significantly different. The majority of the potato arrivals are between the months of June and September in the hills while this is between November and March in the plains. This seasonal difference in production patterns might present significant opportunities for the hill potatoes as they could supply the market during a period of the year when potato production is low in the rest of the state and the country.

These seasonal opportunities are further reflected in Figure II.4. They show the highly seasonal pattern in potato prices in the two most important potato *mandis* of state. During the last year (April 2006-March 2007), potato prices were twice and four times as high in the Haldwani and the Dehradun *mandi* respectively in the June-September period compared to the December–March period. This illustrates the high price premium that hill potatoes can fetch because of harvest timing. While potatoes can be stored and cold storage potatoes do enter the market during that period, consumers are willing to pay a significantly higher price for fresh potatoes. The hills are the only regions that are able to supply the market with fresh potatoes during that period.

Figure II.4 **Seasonal Patterns in Potato Prices in Uttarakhand (April 2006-March 2007)**

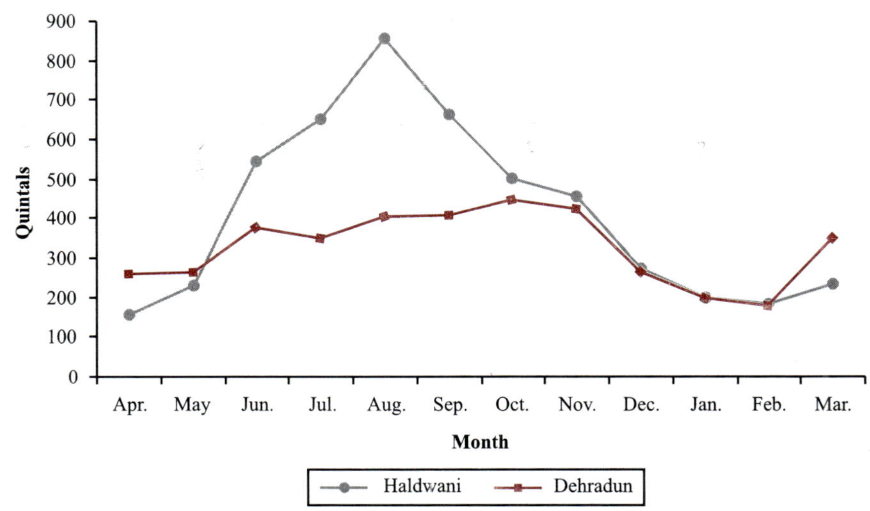

Source: Own calculations based on Agmarknet dat.

We discuss further potato consumption trends and price behavior at the national level as well as in the terminal market of New Delhi. The graph (Figure II.5) shows how potato consumption has evolved over time (between 1990 and 2005). We see a strong increase in per capita consumption, from 35.2 g/day in 1990 to 44.1 g/day in 2005 or an increase of 25 per cent over this 15 year period. The biggest increase has happened in the period between 1993 and 2000. Per capita potato consumption is leveling off in the last five years and is then even slightly on the decrease.

Figure II.5 Potato Consumption in India 1990-2005

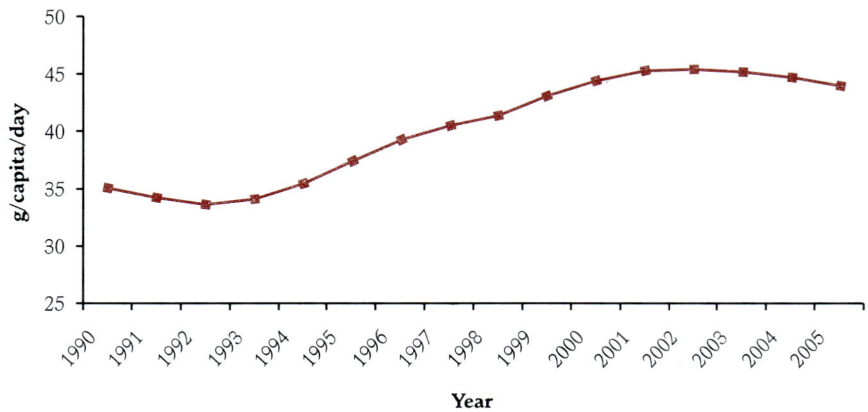

Source: FAOSTAT

Figure II.6 Real Wholesale Potato Prices in New Delhi (March 2003-September 2007)

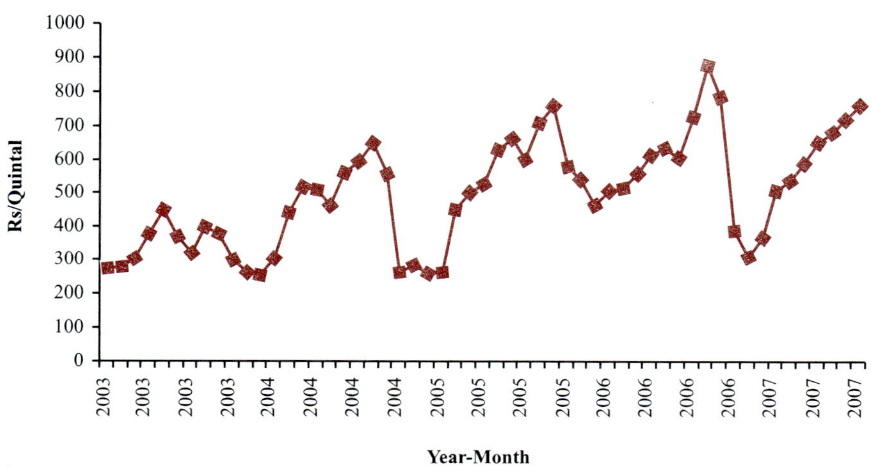

Source: Own calculations from Agmarknet data

Figure II.7 Seasonal Price Movement in Potato Wholesale Prices New Delhi (2003-2007)

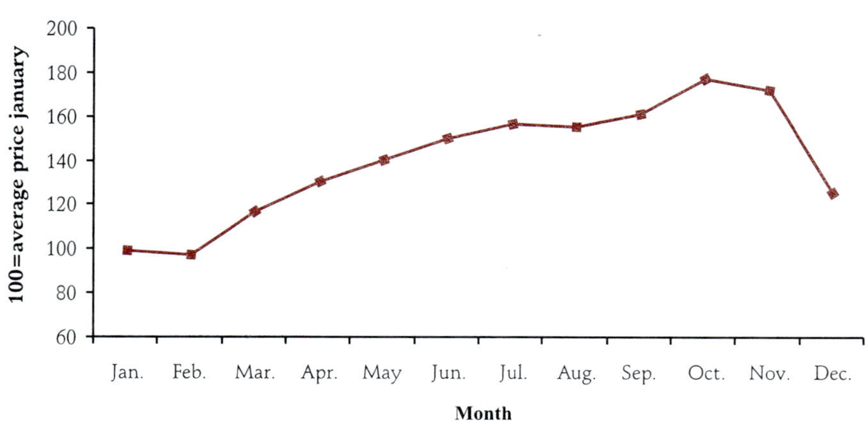

Source: Own calculations from Agmarknet data.

Relying on Agmarknet price data, modal prices recorded at the wholesale *mandis* in New Delhi are presented in Figure II.6. To control for inflation, we deflate all prices with the Wholesale Price Index as published by the Indian Ministry of Commerce. We equate the index equal to 1 in January 2007. All prices are thus expressed in January 2007 prices. We note a strong variability of the potato prices. The potato price shows a strong seasonality and a strong trend upwards over the period 2003 to 2007. When we use a simple regression model to separate trend and seasonality where we regress the log of the price of potatoes on yearly, monthly, and market dummies, we find that the real price in 2007 has increased by 57 per cent compared to the year 2003.

The strong seasonality of potato production as well as the costly storage leads to large seasonal fluctuations in Delhi as shown in Figure II.7. The highest prices are recorded in the months of October and November, the lowest prices in the months of January and February. As the dataset does not make a distinction between fresh and cold storage potatoes but as the majority of the traded potatoes are potatoes from the plains, this seasonal movement reflects the rewards to storage. The rewards of storage include the physical costs of storage (infrastructure costs and losses) as well as the opportunity costs of capital. The average monthly increase amounts to about 8 per cent. It shows a slight plateau in the months of August and September, potentially because of the increasing competition of the hill potatoes that enter the market during that period.

To further understand retail in Delhi, a survey was organised at different retail outlets to evaluate the price differences between modern (including Mother Dairy) and traditional outlets and between cold storage potatoes and hill potatoes. The results are shown in Figures II.8 and II.9. First, the results show different price setting behaviour between modern retail and traditional retail. While there is a significant overlap, the prices that were observed in modern retail were all on the lower end compared to traditional retail. Through better supply management, a reduction in wastage and going around licensing rents, it seems that modern retail is able to benefit consumers. Second, we find a significant difference between the average price of hill potatoes and cold storage potatoes. Hill potatoes do seem to command in the Delhi market a 50 per cent premium (Rs. 5 to 6/Kg) over the cold storage potatoes during the period of the survey (June 2007).

Figure II.8 Potato (Hill) Retail Prices in New Delhi

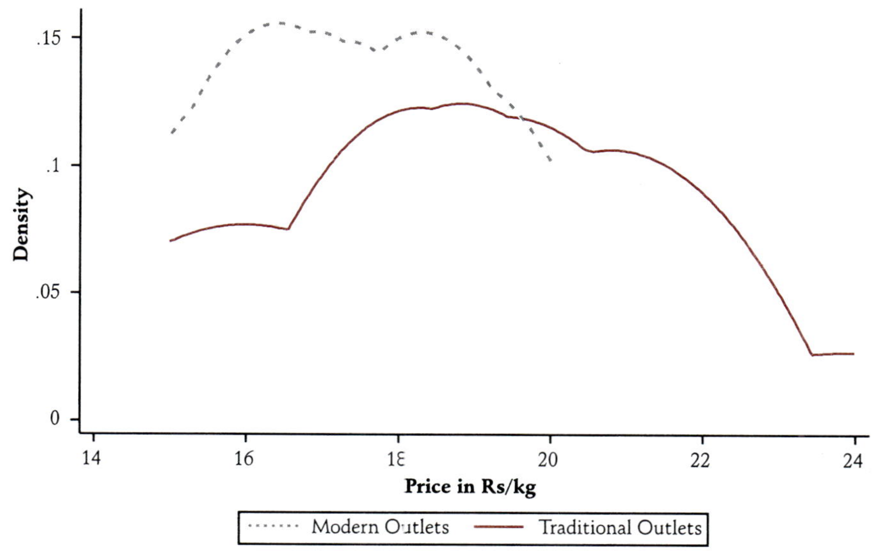

Source: Own retail survey.

Figure II.9 Potato (Cold Storage) Retail Prices in New Delhi

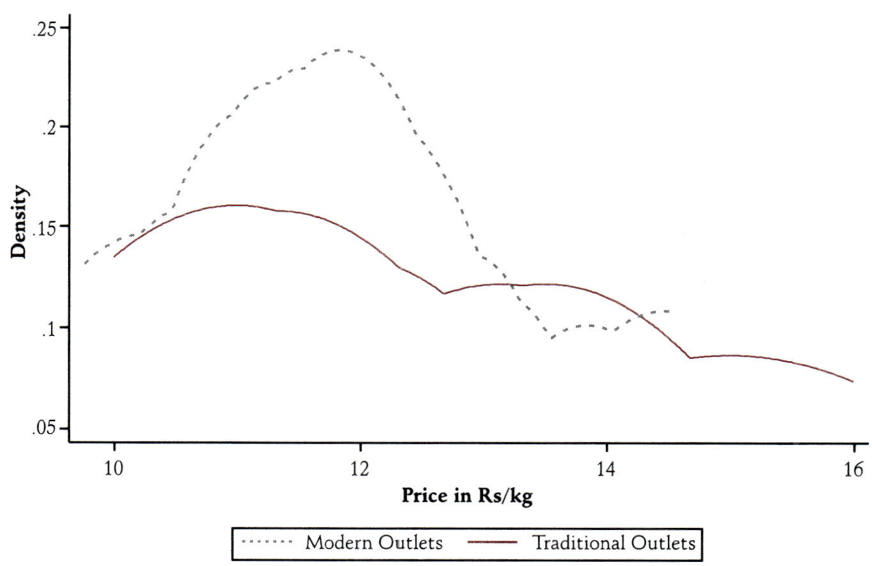

Source: Own retail survey.

During our field visits, we conducted a significant number of interviews with traders as to better understand marketing costs and how they contribute to price formation for the customer and for the producer. Using an average of results coming out of our interviews, we find that the marketing margin makes up Rs. 5.5/kg or 33 per cent of the retail price in the cities of Uttarakhand. This share would probably change for the produce that is marketed in other regions or cities.

A big part of the marketing margin goes to the commission rates and the taxes that are charged. This amounts to 30 per cent of the total marketing margin. The retail margin contributes Rs. 2/Kg or 36 per cent to the total marketing margin between producer and consumer. While this is seemingly high, it is actually significantly less than in the other value chains studied in this report (apples, tomatoes). Given that potatoes can easily be stored, losses are relatively limited and amount on average to 5 per cent of the total production marketed at the farm level.

The transport costs are prone to inaccuracies given that we do not have a database on the average potato producer in the state of Uttarakhand. We use the transport costs as the costs of bringing produce from the important potato producing areas in Bhageswar to the wholesale market of Haldwani, the second most important potato procuring wholesale *mandi* in Uttarakhand. The costs of bringing potato from these villages to the wholesale *mandi* are estimated to be around Rs. 1.12/kg or 20 per cent of the total marketing costs between producer and retailer.

II.5. Conclusions

Potato is an important staple in the food economy of India as well as in the state of Uttarakhand as it is the most important vegetable grown in the state. Its total production in the state amounts to 450,000 tonnes. A big advantage of the potatoes grown in the state of Uttarakhand, especially for the potatoes that are grown in the hills, is their off-seasonal harvest period. While cold storage potatoes are commonly available in the market during that period, consumers are willing to pay a high premium for fresh potatoes given that cold storage potatoes suffer from sugar conversion from starch.

Different interventions could be made to improve the functioning of the value chain. Our simulations show that the largest effects in the value chain could be obtained by interventions in the regulatory environment. The elimination of taxes and de-licencing of agricultural marketing—inducing investments by modern retail—would have large pay-offs and would have few budget implications for the state government.

Agricultural research could contribute enormously in the improvement of productivity. Hybrid seeds are currently not widely available for potatoes that limit an increase in their productivity. The number of research organisations that deal with potato and the specific problems of the hill potato are also limited. Given the impact that research could have for a large number of smallholders, there is a potential large pro-poor impact of agricultural research in this area.

Annex III

Off-season Vegetables:
The Case of Tomatoes

III.1. Methodology

Interviews were held with producers, representatives of farmer groups, consolidators, representatives of Mother Dairy, managers of a local NGO that helps in the organisation of the farmers in the district of Uttarakashi (HARC) and with wholesalers, retailers, commission agents and representatives of the APMC in the wholesale market of Dehradun. We report most of the findings of these interviews in this write-up, starting from the producer level to primary marketing options, organisation of the wholesale markets and retailing.

III.2. Production

III.2.1. The Indian Situation

Tomato is a summer vegetable, with ideal requisite temperatures ranging between 25 and 30 degrees Celsius. An average night temperature of 15 to 20 degrees C augurs well for a good crop, while any extremes of daytime temperature (below 15 and above 40 degrees C) would destroy the crop entirely. In addition, dew and hailstorms are also known to harmfully affect standing crops in the fields. Inadequate irrigation and low moisture content in the air are also potentially damaging. It is recommended by scientists that tomato should not be planted on the same piece of land for two seasons in succession; further it is warned that planting tomato on plots that have very recently grown aubergine and potato crops would enhance the chances of pest and disease attack. Tomatoes are often grown on land that has been used for wheat cultivation in the main season.

Tomato production in India has increased steadily over the last 15 years (Figure III.1). The production has increased by a high 86 per cent in 2005 compared to 1990. The area allocated to tomato production saw a similar increase over the period, 87 per cent, indicating that yields have changed little over that period.

Figure III.1 **Trends in Tomato Production in India 1990-2005**

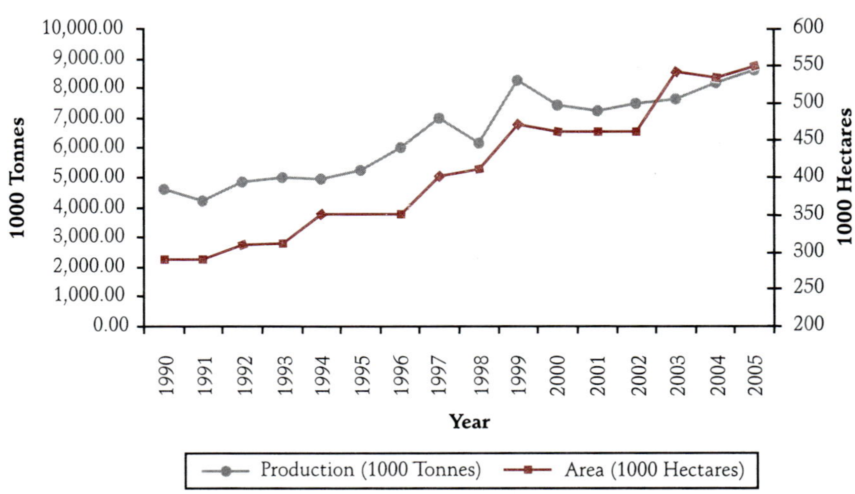

Source: Faostat

III.2.2. The Situation in Uttarakhand

The agro-climatic conditions of the hill states of the Himalayas are favourable to the cultivation of off-season vegetables, with tomato being one of the largest such income generating crop. The cultivation of tomatoes in the hills, possible only in months during which the plains are unable to produce large quantities, makes it a unique off-seasonal crop with large demand from the *mandis* of the plains. The sowing seasons are, in the plains from October to February and in the hills from February to June. Traditionally, tomatoes in the hills are ready for sale in the period from June to July. Traders estimate that tomato trade has increased during that period by 10 per cent to 20 per cent per year in recent years in the state of Uttarakhand. Most of the tomatoes during that period are produced in Himachal Pradesh, Maharasthra and Uttarakhand. The boom in the hills of Uttarakhand only happened in the last 7 years.[4]

The majority of producers use hybrid seeds as well as fertilisers and pesticides. Inputs can be obtained through farmers' federations as well as on private markets. Input prices acquired through the federations are about 30 per cent or 40 per cent lower than on the market. Seed companies are able to give large discounts to the federation because of the large quantities that can be negotiated at once. Each federation usually acquires around 25 Kgs of seeds. The value of 1 Kg of seeds amounts to Rs. 200.000. These are then being sold to farmers at Rs. 200-400 for 10 grams. The government also sells seeds but those are typically non-hybrid seeds and often of lower quality. The seeds are provided through KVKs and the department offices at the block level of the Ministry of Horticulture or Agriculture.

Pesticides might be significantly over-used as farmers stated that they would spray about once every week or 10 days.[5] Most farmers believe that spraying will affect the quantity—but not the quality—produced. One of the most commonly used pesticides is Nu wan. Pesticide use in tomato cultivation is a common practice in the hills.[6] Farmers state that not using pesticides would lead to yield reductions between 25 per cent to 30 per cent and to the production of smaller-sized tomatoes. All the same farmers also feel that organic tomatoes are tastier. However, there is currently no buyer willing to pay a premium for organically produced tomatoes. Only one large farm in the area is able to market organic tomatoes and receive a premium for it.[7] Fertilisers are also commonly used on tomatoes.[8]

A major production problem of tomato cultivation—or any agricultural cultivation—in this area is the destruction by wildlife, especially monkeys. Some farmers estimated the losses to more than 50 per cent in some years. Another production problem of perishable crops such as tomatoes is the high price variation. Farmers stated that producer prices varied last year between Re. 0.50 and Rs. 20 per Kg. If the farmer were to sell the produce during the glut period, he would make little profit. On the other hand, if the farmer happened to sell during the peak price period, he would benefit from large windfalls. It seems that little can be done by producers as well as the government about these movements. Better storage conditions and varieties with longer shelf life may be better suited to help smoothen out daily price variations.

As to better capture market demand and to avoid gluts, there is an incentive to spread out the harvest periods over more months. Through technology, adoption of better suited varieties and better coordination between different producing regions, the farmer of Uttarakhand might be better able to capture the market demand. Interviews with representatives of HARC, an NGO that closely works with these farmers show that they are actively trying to spread out the sales of tomatoes from June through October in the area that was visited. However, while there is little outside competition with the tomatoes produced during June and July, competition for markets from tomato producing areas such as Maharashtra (Nasik) and Bangalore kicks in from August onwards.

III.3. Marketing Channels

Tomatoes are sold through different channels. For farmers who are members of a farmer federation, it is estimated that about 50 per cent is sold through these federations. The most important buyer of this federation is Mother Dairy (MD) that sells this produce through its SAFAL outlets in Delhi. The federations also sell

---

4. The tomato production was taken away from areas around the city of Dehradun, Saharanpur and other places in Uttar Pradesh.

5. Unfortunately, we have no data on level or dilution of spray.

6. Pesticides are subsidised up to 50 per cent by the government. This is also the case for 'organic' pesticides. In case of widespread disease outbreaks, the government sometimes even provides them for free.

7. Maharashi Orchards. They are reportedly able to sell organic tomatoes in the Delhi market for Rs. 40 to 50 per Kg. However, this is a large farm that also produces organic apples and is the only one that we heard of in the area that was able to obtain a premium for organic tomatoes.

8. Fertiliser prices are also subsidised by the government (up to 50 per cent). Market prices in the area are for a bag of 50 Kgs for Urea: Rs 250; Dap: Rs 450; Mop: Rs 350.

to other buyers who have been carefully selected through the federations as to assure that the contracts are respected, prices are high and payments are regular. 50 per cent of the produce of the members is sold individually through traditional markets. For non-members, these individual transactions amount to 100 per cent of all production.

Different qualities of tomato are sold to different market channels. The best quality tomato (A-quality) is bought by the modern market channel, Mother Dairy. The best quality tomato at the producer levels is defined as a tomato of a size of 45-60 mm diameter, 80 to 100 grams and a maximum of 25 per cent of it is red (i.e. 1/4th of the lot is considered ripe; this requirement ensures that tomatoes are not overripe at the final destination and in retail outlets). To achieve the high quality tomato, the major requirement is the right time of picking. Traditional market channels buy good as well as bad quality. The rewards show up at the auction but it is estimated that for the same quality that is delivered to MD, farmers would receive a significantly lower price in traditional channels.[9]

Mother Dairy was set-up as a national dairy cooperative in 1972 for the procurement of milk. It diversified in the mid 80s into the sale of fruits and vegetables that it is selling in a large number of outlets in Delhi. It is estimated by the representative in Uttarakhand that Mother Dairy procures 200 trucks of produce a day and that it sells 600 tonnes of produce daily through their different Delhi booths. Most of the produce that it sells is procured directly from farmers. Only rarely does MD procure produce from *mandis*.

Mother Dairy collects the off-season vegetables directly from federations that they may have started up themselves or which had been started up by organisations such as HARC. With the help of the federations, they have set up collections centres for the four different federations in the Naugaon area where different trucks collect the produce daily. Produce is mainly procured from within a radius of 1 Km of the collection centre. Mother Dairy procured 700 tonnes of tomatoes from all the federations as a whole last year, i.e., about 20 tonnes daily from all the collection centres in Uttarakhand combined. Yearly targets of the production of tomatoes are set in joint meetings between the federations and Mother Dairy a significant period before the actual production date. The target for the current year is 1000 tonnes. For transportation, it uses trucks that can load from 4 to 7 tonnes. MD currently only buys tomatoes and other off-season vegetables during a two-month period, from June through July. Tomato prices during this period usually range between Rs. 5 and 8 per Kg.

The obligatory 2.5 per cent *mandi* tax on agricultural produce transactions is initially paid by the federation but is later reimbursed by MD. MD focuses on selling high quality vegetables and produce. However, they do not procure organic produce and do not pay a premium for it. At the time of our survey, MD procured tomatoes at Rs. 8.25 / Kg and was selling them at Rs. 14 /Kg in their booths in Delhi. Farmers pay Re. 1 for every crate delivered to the federation. This money is being used towards a common fund managed by the federation to help farmers deal with losses and unforeseeable damage, acting thereby as an insurance fund.

With the help of the Himalayan Action Research Centre (HARC) and after a number of market studies in different *mandis* had been conducted, the federations identified a series of traders in different markets who were willing to follow the stringent requirements of the federations. These include the fixing of procurement targets, upfront payments and payments through bank accounts. About 16 traders have been registered as such. It is estimated that of the tomato produce that is being sold to the reliable wholesalers (including MD), 40 per cent is going directly to Delhi, 30 per cent is going through Vikasnagar, 10 per cent is being sold in Dehradun and 20 per cent in other markets such as Meerut and Saharanpur.

Tomatoes that are not being sold through the federations are being sold through individual commission agents. Commission agents rely on truck drivers or on consolidators to collect the produce. Crates are left in the morning at those places near the border of the road where commission agents are expecting produce from. Note that these places are often already far from villages. For example, during our visit in the area we noticed

9. Rewards for quality tomatoes exist, even on traditional markets. However, they differ regionally. There is a large demand and reward for the A-grade from the New Delhi market. B-grades usually sell well in second-tier cities such as Dehradun, Meerut and Lucknow. No or little rewards exist in small *mandi* towns such as Saharanpur and Muzaffarnagar.

that in one case villagers had to carry produce on their backs for 2 or 3 hours before reaching the designated point. In another case, produce was transported on the back of donkeys at the cost of Re 0.50 /Kg, or about half the cost of bringing the produce from Naugaon to Dehradun for example.

The consolidator, often somebody who lives in the area, organises the collection from farmers (about a dozen a day). However, he does not negotiate prices or arrange payments. He assures that crates are properly filled (a slip is added to each crate as to assure that no confusion exist on the ultimate CA) and that produce is collected at the right spot on the road.[10] He coordinates by phone with truck drivers and with the CA that he works with.[11] He is usually paid Rs. 5 per crate to do this. It might be possible that the CA—through the consolidator—provides inputs to the farmer (seeds, fertiliser, pesticides) in which case the farmer is locked into a contract with this CA. This usually leads to a lower output price. Otherwise, farmers might choose an available CA the day of the pick-up. Farmers are not aware of the price that their produce will fetch at the day of delivery. A payment slip would be delivered to the farmer typically two days after the sale took place. The payment slip would indicate the price that was obtained for the crate at the auction, the transport costs to bring the crate from the side of the road to the mandi, the commission rate (usually 6 per cent but it might go up to 8 per cent) and other costs such as telephone expenses incurred. The farmer will be paid the net price, deducting these different costs mentioned.

While farmers have the right to claim their payments immediately, they usually do so only once or twice in the season to avoid paying for transport costs and to reduce the lost opportunity costs of travel. The farmer in this area is typically not (physically) present during the transportation and at the auction and there is no assessment of the quality and the quantity of the produce at the time of delivery. Some traders might thus behave opportunistically by claiming different weights than obtained, claim higher losses (losses are the responsibility of the farmer until the produce reaches the wholesale market), different prices, and different transport costs than paid in reality.

Discussion with focus groups of farmers indicated that the prices that were obtained by the farmers surprisingly did not depend on the quality of the (non-damaged) produce but were dependent on characteristics of the farmer, including the level of advances received, the education of the farmer, the quantities sold and the level of information of the farmer.[12] It seems that access to mobile phones in the area—which became available only one year ago—has especially increased the information available to farmers and that farmers are now better able to play off different CAs against each other. Focus groups of farmers indicated that around 50 to 60 CAs were active in the area, indicating an apparent healthy level of competition.

Farmers state various benefits from selling to MD:

1/ *The pricing mechanism is transparent.* MD uses the Azadpur (New Delhi) wholesale price for tomatoes on the day of delivery at the collection center as the reference price and subtracts the transport costs from this. Transport costs are valued at Rs. 2/Kg from Naugaon to Delhi. This transport cost is significantly lower than transport costs charged by private transporters. MD can offer more competitive rates as they are able to negotiate lower cost contracts with transporters as they offer them contracts for a whole year (as they procure in other periods from other states). Transporters who work for Commission Agents (CA) are only able to work seasonally in the tomato trade and seemingly charge higher transport prices because of slack seasons.

2/ *Producer prices are higher.* Members of the federations state that the prices that MD pays are, on average, Rs 1 to 2 /Kg higher than those offered by private traders. At the time of our survey, the prices offered by MD were Rs. 8.25 /Kg. This compared to a net price of about Rs. 5.60/Kg offered by private traders.

---

10. Tomatoes used to be collected in wooden crates. They were more costly than plastic crates and could not be used for a long time and would lead to higher wastage during transportation given the difficulty of right stapling. This led to an important part in the marketing costs. Wooden crates have now been universally replaced by plastic crates, which are owned by the CA. Good plastic crates sell at about Rs. 170 and lower quality ones at Rs. 110 They can be used for about 1 to 2 years and are thus seemingly a minor cost in the marketing chain.

11. A CA in Dehradun said that he had between 10 and 12 consolidators that do this work for him.

12. Farmer groups estimated the price difference between a lot of 100kgs *versus* 1000kgs at Rs. 0.5/kg to Rs. 1/kg. The costs of stay in the city might also be covered for bigger farmers by the CAs.

3/ *Prices are the same for everybody.* MD gives the same price for everybody no matter if the produce is delivered in large or small lots and regardless of the background of the producer. Producers feel that private traders would offer lower prices to poorer and less-informed farmers and to farmers who sell smaller lots.

4/ *There is no cheating with weights.* Crates are standardised for MD and there is little discussion on the weights of a crate. Farmers who sell through the traditional channel complain regularly about being cheated in weights as no scales are present at the loading places on the roadside.

5/ *Payments are regular.* MD pays by cheque through the federation and farmers receive their payment in less than a week after the delivery has been done (federations deposit cheques in bank accounts of the members). While farmers would be able to collect the money from the CA the day after the transaction was done, farmers only travel to the *mandi* once or twice a season in practice given the high transport costs and the opportunity time of travel to the big cities such as Dehradun or Delhi.[13] While some of the farmers are paid at the farm gate, it is estimated that about 80 per cent of the farmers are paid at the *mandi*.

6/ *Bank payments avoid intra-household allocation problems.* An additional advantage, voiced by women in particular, is that payments made into bank accounts give men less opportunity to spend the money on the *mandi* or near the *mandi* (women rarely go the *mandi* as this is the responsibility of the men) on goods and services that are of little use to the household. Women estimate that less than 50 per cent of the earning of produce sales would actually make it for household necessities if farmers are paid at the *mandi*.

However, there are also several disadvantages. They include:

1/ *MD only purchases the good quality produce.* MD increasingly insists on good quality and this year it has stationed an employee at the collection centers as to assure that only the good quality produce is loaded and transported. Last year, about 50 per cent of the produce was rejected on quality grounds and prices paid for these lots are about Rs. 3 /Kg lesser.[14] There has been much discussion on these quality standards. Some farmers feel that MD is not consistent in its quality standards as it is more lenient in periods of scarcity than in periods of glut. Unlike with MD, at the *mandis* the farmers are assured that all qualities of produce can be sold, albeit better quality may command a higher price even in the *mandis*.

2/ *MD does not provide input advances.* Commission Agents (CA) are willing to provide input advances to the poorest farmers as to potentially lock them into contracts with themselves. Focus groups indicated that about 20 per cent of the farmers would use input advances from the CAs. Almost all of them could obtain these input advances in case of urgency. However, while there are usually no interest charges applied to input advances, there are complaints about the prices that are being charged to inputs. The CAs usually charge the unsubsidised rates (the Maximum Retail Price, MRP) for inputs while most of these produce can in reality be obtained with a 50 per cent subsidy. MD on the other hand does not give input advances. The federations where MD procures from might however facilitate access to credit for their farmers.

3/ *There is only a limited quantity that can be sold to MD.* MD fixes a daily target of tomatoes that it wants to procure. While it may slightly exceed it for particular days, regular communications with the headquarters in Delhi will ensure that procurements are in line with the demand for tomatoes from the stores in Delhi. Excess production would therefore not be procured.

III.4. Wholesale, Retail and Demand

Tomato production is characterised by a large seasonality (Figure III.2). While off-seasonal tomatoes (July-September) are important in hill production, the majority of the tomato trade takes however place in the period between December and March when tomatoes in the plains are produced (in Uttarakhand as well

13. Personal travel costs to Dehradun are only about Rs. 120 per person. However, it might take up to six hours one-way to travel there.

14. MD sells the rejected produce in Delhi on the Azadpur *mandi*; the reason for rejection is usually related to the size of the tomato. 90 per cent of the rejection is reported to be linked to the size of the tomato. 10 per cent is due to other factors such as diseases, colour, or being overripe.

Figure III.2 Tomato Arrivals in *Mandis* in Uttarakhand (April 2006-March 2007)

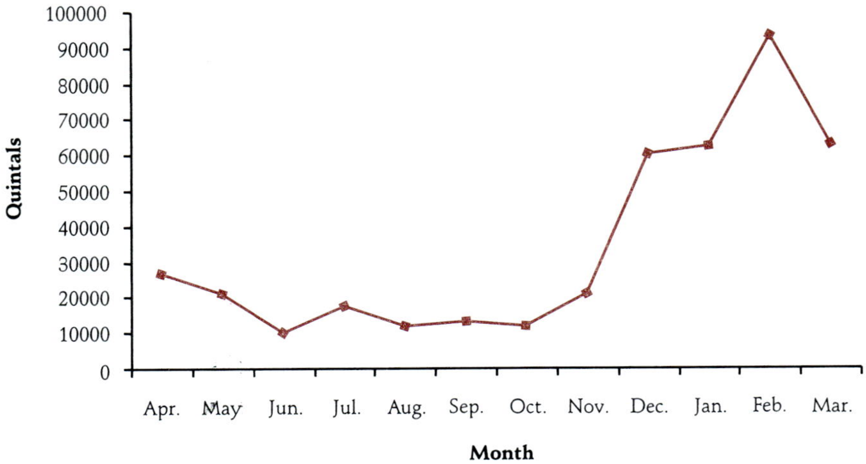

Source: Own calculations based on Agmarknet data.

Figure III.3 Tomato Consumption in India between 1990 and 2005

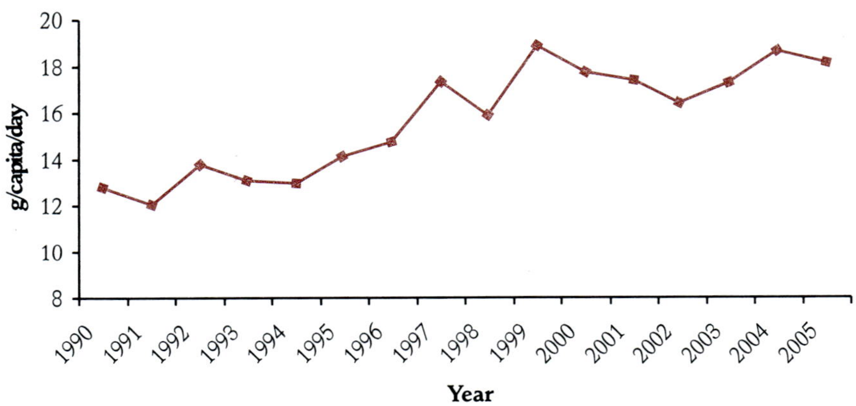

Source: Faostat

Figure III.4 Tomato Wholesale Price Index 1993-2006

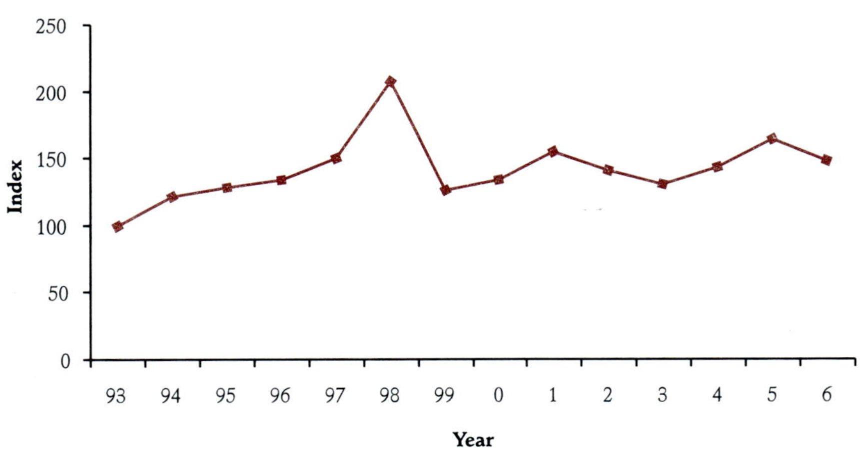

Source: Office of Economic Advisor, Ministry of Commerce.

as in the rest of India). The larger production also leads to lower prices during that period as we will see later.

The commission agents that are active on these markets are subject to the APMC Act that is currently in effect in Uttarakhand. The APMC Act states that economic actors in the agricultural value chains are subject to licensing by the APMC. Any commission agent, transporter, wholesaler and processor who deals in food trade has to obtain licenses from the APMC. Licences are given out for a minimum of one year and a maximum of five. It is estimated that about 50 per cent of the tomatoes produced in the western part of the state are sold directly to wholesalers and commission agents of Delhi. The rest of the tomatoes goes largely though the wholesale market of Dehradun and Vikasnagar from where it may be dispatched to other wholesale markets (such as Meerut, Saharanpur, etc) or to retailers in Dehradun.

Figure III.3 shows the evolution of tomato consumption in India over the last 15 years. Consumption is growing very rapidly. Daily per capita consumption has increased from 12.7 gram per capita in 1990 to 18 grams per capita in 2005, an increase of 42 per cent. Most of the increase has happened in the 90s though and consumption in the 2000s has been stable.

Figure III.4 shows the recent evolution of the tomato price. The price in 2006 is 50 per cent higher than the price in 1993. However, most of this change has happened between 1993 and 1997. The tomato price has stayed at the same level since.

Tomato prices are further characterised by significant seasonality. Figure III.5 shows the seasonal evolution in a typical year, April 2006–March 2007, in the wholesale markets of New Delhi and Haldwani. The graph shows how prices increase significantly in the period between June and October, the off-season, in Delhi as well as Haldwani. The graph also shows that the wholesale price in Haldwani always stays about Rs. 200 per quintal lower than the price in Delhi, roughly reflecting the transaction costs between these two markets. It indicates that tomatoes from Uttarakhand could be profitably supplied to the Delhi market throughout the year.

Figure III.5 **Tomato Prices on Wholesale Markets in Uttarakhand and New Delhi (April 2006-March 2007)**

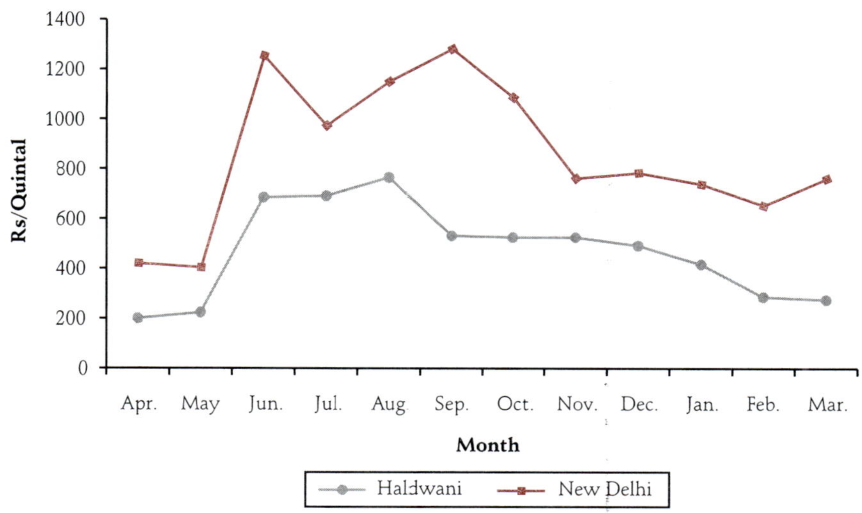

Source: Own calculations based on Agmarknet.

Current food retailing in Dehradun relies largely on informal and small traders. Different types of traditional outlets exist: 1/ the push cart vendor; 2/ the market vendor; and 3/ the shop vendor. All retailers buy their produce on the wholesale market in Dehradun. Some buy it through auctions while others buy it directly through wholesalers or retailers on the *mandi*.

There is currently no modern food retail in Uttarakhand. As to better understand the impact of modern retail on pricing, a small survey was organised in Delhi at different retail outlets to evaluate the price differences between modern (including Mother Dairy) and traditional outlets for tomatoes. The results are shown in Figure III.6. While there is a significant overlap, the prices that were observed in modern retail were all on the lower end compared to traditional retail. Through better supply management, a reduction in wastage and going around licensing rents, it seems that modern retail is able to benefit consumers.

Figure III.6 **Tomato Retail Prices in New Delhi**

Source: Retail survey

Traders were also talked to in the Delhi markets as to better understand the dynamics that are going on in the terminal retail market. 40 to 50 trucks of tomatoes arrive every day in the Delhi market. The origin of the traded tomatoes changes dramatically over the year. Quality is rewarded, even in traditional retail markets in Delhi. Price differences are mostly based on the size of the tomatoes. Larger as well as small tomatoes fetch often a significantly lower price. The difference with the medium-size tomato is up to 20 per cent. Green, small tomatoes obtain a much lower price (-40 per cent). Controlling for size, the variety Hem Sona, grown in the hills of Uttarakhand and H.P., is valued 15 per cent to 20 per cent higher than the hybrid tomato. The variety Hem Sona is stated to be preferred because it is less vulnerable to fungus attacks and because it has a longer shelf life. This variety has thus over time become more popular. However, in contrast with hybrid tomatoes, it can only be grown once a year.

III.5. Conclusions

Different types of interventions could be envisioned towards a better structured tomato value chain. At the grower level, the introduction of different varieties might sustain and improve the tomato boom in the state of Uttarakhand. First, given the availability of different varieties and the strong variation in climates due to differences in altitudes, other varieties could be tested to exploit these opportunities and thus be linked with the market for longer periods instead of the current two months. It also seems that a high pay-off could be achieved through the introduction of varieties that are better suited for transportation, storage and a longer shelf life or through post-harvest technologies, such as sheds along the road, which would reduce the exposure of tomatoes to the sun. This would increase shelf life and reduce wastage in the marketing system leading to lower margins, higher price for producers, lower prices for consumers, larger quantities sold and improved incentives for consumption and production. One necessary condition for the production of tomatoes and off-season vegetables is access to irrigation and it seems that the large increase in some parts of the state would not have happened if access to this were not sufficiently available.

Different interventions could be made as to improve the current marketing system. Our findings reveal that legislation of the APMC is largely outdated and is moreover not respected by actors within the value chain. To move to a modern marketing system, it is of utmost importance that current regulations be modified. The APMC Act that is currently in being was designed in an era when the Indian agricultural economy was facing significantly different problems than is currently the case. A move from a tax-based[15] regulated market system to a trade facilitating and competition enhancing system would be to the benefit of consumers and producers alike.

Finally, while there are currently only few modern retailers active in the state of Uttarakhand, their increased presence might mean a significant boost to the agricultural economy and it should be a priority of the Government of Uttarakhand to further facilitate their investments. Our case study of procurement systems of Mother Dairy shows that this is largely beneficial for tomato farmers and that prices offered to farmers are significantly higher than through traditional marketing channels.

15. Where produce that is exported outside the state is taxed twice, i.e., within the state and in the receiving state.

Annex IV

Organic Farming

IV.1. Introduction

Recent years have seen a heightened concern for food safety, quality and ecological concerns in food markets across the world. This has translated into a noticeable shift in consumer preferences towards chemical-free and non-synthetic products, particularly in developed countries. Food items labeled "biological", "ecological", "bio-dynamic" and "organic" all fit this bill. Arguably the largest and most popular category amongst these is that of Organic Agriculture, which from our perspective, would represent not only the label that a food product may have, but also the entire value chain and production system it represents. The International Federation of Organic Agriculture Movements (IFOAM) describes organic agriculture as follows (taken from UNCTAD, 2004):

"Organic agriculture includes all agricultural systems that promote the environmentally, socially and economically sound production of food and fibres. These systems take local soil fertility as a key to successful production. By respecting the natural capacity of plants, animals and the landscape, it aims to optimize quality in all aspects of agriculture and the environment. Organic agriculture dramatically reduces external inputs by refraining from the use of chemo-synthetic fertilisers, pesticides, and pharmaceuticals. Instead it allows the powerful laws of nature to increase both agricultural yields and disease resistance."

In practice, food products may be labeled "Organic" once it is verified that they have been produced under specific production processes which enable it to meet specified standards. Such verification is conducted by authorised certification agencies.

IV.2. Global Trends in the Organic Food Sector

According to the UNCTAD (2004) an estimated 22 million hectares of land are under certified organic cultivation the world over. Nearly half of this area lies in Australia. The next large organic producers include the USA, Germany, Austria, Italy and Argentina. In terms of organic certified land as a percentage of total agricultural acreage, the highest percentages of such land are in Austria, Switzerland, Italy and Scandinavian countries. In most developing countries, the share of organic area is less than 0.1 per cent of the total cultivated area. Organic farming is practiced in 110 countries and its share of farmland continues to grow.

The organic food market shows much promise. According to the EXIM Bank (Agri-Export Advantage, 2006) while the conventional (non-organic) food industry has grown by 2-3 per cent over the past few years, the organic food industry has grown to the tune of 17–22 per cent. Between 2003 and 2006 alone, global organic food sales jumped from US$ 25 billion to US$ 31 billion, and are projected to reach US$ 102 billion by 2020. The major markets for organic food products are the USA, the EU and Japan.

In the EU the market saw a formidable spurt in trade over the nineties, which recorded 20–40 per cent annual growth rates over this time. Of late, growth rates seem to have come down, and the markets have tableaued. The largest players, Germany and the UK together, account for 43 per cent of the EU market. The EU market displays considerable variation in terms of popularity and stage of development of organic food markets. For instance while in Austria, Germany, Switzerland and the UK, organic produce is sold by leading supermarket chains in countries such as the Czech Republic, Greece, Spain and Portugal the trade is small and not very organised. Specialised organic food stores and direct sales dominate in countries such as Italy, the Netherlands, France and Belgium. The resounding trend however is that though lower than before, growth rates of organic food products in European markets are positive. Increasing health concerns and the simultaneous positioning of organic foods as a niche product in the face of rising incomes have sustained the markets in recent years. Generally almost all organic crops command a premium over conventional crops in EU markets, roughly between 15–20 per cent, though the margins vary across crops and countries. Germany offers the highest premiums, while Sweden the lowest.

Although currently small in comparison to the US and the EU, Japanese markets are expected to witness growth in imports of organic crops of about 40 per cent according to the Japanese Integrated Market Institute (Agri-Export Advantage, 2006). Other markets likely to emerge in the years to come include Saudi Arabia, UAE and South Africa. In the US, the annual market is estimated to be worth anywhere between US$ 16.3–US$ 29.7 billion. For 2005, retail sales were estimated at US$ 12.8 billion. The US meets a part of its organic food requirements through imports, which in 2002 were to the tune of US$ 1.5 billion.

This clearly indicates a considerable demand in world markets for organic produce. Coupled with a small but steadily expanding domestic market (current consumption is estimated at 1200 tonnes annually, potential customers at 2-3 million), this augurs well for the possibility of organic food products from Uttarakhand finding a niche market and ready buyers.

IV.3. The Uttarakhand Context

Organic farming is widely believed to be, not just a possible growth path to be followed by the state, but also a solution to many economic and ecological problems of the state. In the foothills and plains in the districts of Dehradun and Udham Singh Nagar, agriculture practiced is largely based on heavy chemical use in the form of fertilisers and pesticides. This has resulted in heavy damages to water, soil and also to farmers' health. Some of the crops cultivated here, particularly Basmati rice have much potential to be profitably cultivated organically, and thus escape the harmful effects of chemical-based production.

The case of the hill districts is rather different. Much of this area is either unsuitable for farming, covered by forests or too small in size to allow large scale farming. For instance, in Uttarkashi and Chamoli districts, percentages of land available for farming of total land stand at 4 per cent and 5 per cent respectively (Alam and Verma, 2006). Coupled with this is the issue of poor irrigation in the state. For example, it is estimated that a mere 5 per cent of cultivable land in Pauri district has access to irrigation. More recently the state has witnessed drying up of perennial water sources in the upper reaches in the hills; particularly in Almora district (Kumar, 2006). Under these circumstances, most farmers in the hills practice traditional farming methods with low adoption of hybrid, High Yielding Varieties of seeds and limited recourse to fertiliser use. Even where fertilisers are used, the quantities are much lower than in the rest of the country. In such a context of near-subsistence farming and low chemical intensity, the conversion of plots to certified organic will not be accompanied by very drastic declines in productivity. Farmers can therefore be incentivised to switch to certified organic production without as much engagement, persuasion and provision of insurance to mitigate risk as would possibly be the case in chemical-intensive farms.

Organic farming therefore holds promise on multiple fronts. First, due to the prevailing premium in world and domestic markets, organic crops can provide the farmers with higher prices for their produce if marketing channels are strong. Secondly, the decreased use of chemical fertilisers and pesticides would have beneficial impacts on the local ecosystems in terms of reducing soil and water (including groundwater) pollution. An important argument for the propagation of organic agriculture is the reduction in production costs. Organic farmers do not use chemical fertilisers, pesticides, HYV seeds etc. and make substantial savings on expenditures on these (also monetary and non-monetary costs borne due to exposure and proximity to chemicals in the form of poor health). Instead, the need for bio-fertilisers and pesticides increases the economic returns to cattle, and could therefore also encourage people to take up dairying as a supplementary activity. These of course are the potential, or promised benefits. The true impact of propagating organic farming depends much upon the policy environment and how it allows farmers to harness the potential.

We now turn to the institutional and non-institutional channels that are operating in Uttarakhand for the promotion and propagation of organic farming, and seek to identify the efficacy or otherwise of their interventions. Recognising the benefits of organic farming, and seeking to accord it an important position in the state's development path, the Government of Uttarakhand declared itself an "Organic State" soon after it came into existence in 2001. This was coupled with a number of institutional decisions and interventions in the state to promote organic farming practices.

The current policy on organic farming in the state owes its origin to the experiences under the World Bank supported Diversified Agricultural Support Project, i.e., DASP (see Tolia, 2007). This was a project aimed at increasing rural communities' livelihoods options through agricultural diversification, and was implemented in the districts of Dehradun, Uttarkashi, Almora, Nainital and Udham Singh Nagar. The DASP project had a small component called Bio-Village Activity which promoted organic agriculture. The successful results of this activity called for emulation and scaling up with the support of the Ministry of Rural Development, GoI. Young men and women from Below Poverty Line (BPL) families were selected and trained to offer extension services to the larger numbers of poor farmers on organic farming practices such as Bio Dynamic (BD), Effective Microorganisms (EM), Vermiculture, Integrated Pest Management (IPM) etc. This facilitated a significant technology transfer to poor communities with scarce resources. Over the next two years, the number of such Master Trainers (MTs) increased and penetrated each of the 95 development blocks. Presently, the state has 667 bio-villages. The State government floated a proposal for establishment of a Centre of Excellence for Organic Farming, to be called Centre for Organic Farming (COF). Upon securing funding for the same from the Sir Ratan Tata Trust, the COF became operational and was entrusted with the responsibility of providing technical assistance and innovation for organic farming.

In May 2003, the Uttaranchal Organic Commodities Board (UOCB) was established for the purpose of preparing projects, installation of Internal Control System for certification, training and providing support for marketing, product development etc. In recent years the UOCB has emerged as a pioneer institution of sorts, at the forefront of policy and practice in organic agriculture, conducting training programmes, workshops and sharing experiences in national and Himalayan states' fora. The most important role played by the UOCB is the help they provide for obtaining certification. The Board works on the principle of an Internal Control System (ICS) by which the local community is made accountable to ascertain whether or not organic procedures have been followed, tapping in on in-built accountability instead of over-reliance on external inspectors. In addition, the following initiatives have been taken by the state government to promote organic farming (adapted from Alam and Verma, 2006 and Tolia, 2007):

1. Uttaranchal State Seed Certification Agency reconstituted as Uttaranchal State Seed and Organic Production Certification Agency. Entrusting the nodal seed agency with the responsibility of training for organic certification is indicative of the importance the GoUA accords to organic farming. The Agency is the first such State Certification Agency in India that has an ISO 65 certificate.

2. All state-owned farms are being converted into organic farms, with the aim of developing them as centres for excellence for demonstration of effective organic cultivation of select crops.

3. All agricultural research institutions such as the VPKAS and the DRDA in the state have been directed to integrate organic farming into their research agenda and cover various aspects of organic farming.

4. Directive issued to train all staff in agriculture and horticulture departments in organic farming.

5. Provision for separate shops for organic farmers in state-run *mandis*.

These government initiatives/directives very clearly reflect the state's conviction that organic crops cultivation are in the larger interests of Uttarakhand's growth and development. We now turn our attention to some of the initiatives taken up by both private and government agencies in the field of organic farming to draw lessons and understand better the prospects of organic agriculture in the state.

IV.4. Case Studies

We first look at two cases involving cultivation of organic basmati rice in the state; one carried out by Sunstar Overseas Ltd., a private company engaged in basmati rice exports since 2001, and the other run and organised by the Uttarakhand Organic Commodities' Board (UOCB). The results and insights here draw heavily from a comparative study conducted by Alam (2007). Thereafter we discuss the initiatives of INHERE, a local NGO working in Almora district on organic farming.

IV.4.1. Sunstar Overseas Ltd.

Sunstar Overseas Ltd. started operations in the Terai belt of Uttarakhand in 2001, known to be home to the basmati variety. In 2001, Sunstar worked with 100 farmers representing 500 hectares. This has increased fivefold as the number of farmers working with Sunstar now is above 500. These farmers have formed a federation called the "Federation of Small Farmers of Khaddar Area". Sunstar works with both small and large farmers, and the landholdings range between 0.24 and 16 hectares. Sunstar buys up specified volumes of organic basmati rice from farmers and markets it to domestic and exports markets, including to chains in Switzerland and France where the rice is sold as "Fair Trade". Sunstar obtained the "Fair Trade" label in 2004.

Sunstar provides the farmers (through the Federation) with inputs, technical and certification support, quality specifications and marketing support by entering into contracts with them over a period of five years. These are explicit contracts with responsibilities of both parties, and implications of non-compliance clearly spelt out.

To maintain a steady supply of good quality output, Sunstar has developed a technical support package for the farmers. Farmers are provided with extension services so as to train them on best processes to be followed for good quality yield. They have also been trained to make compost from cow dung, and in the use and application of bio-fertilisers and green manure; as well as for bio-pesticides.

Certification is crucial for commanding premiums for organic produce in world as well as domestic markets. However, the costs of certification, especially in the initial take-off stages may be prohibitively expensive in some cases. Sunstar has facilitated organic certification for its farmers. This is provided by SGS, and since 2004 by ECOCERT as well (the latter providing additional certification for Bio-Dynamic farming). The introduction of the more cost-effective ICS has helped the costs to the minimum possible.

Sunstar is one of India's leading basmati rice trading companies.[16] Having marketed 16,000 MT of rice in 2005-06 (15 per cent of this is sold[17] in domestic markets), it recorded a turnover of over Rs. 3200 Million. In world markets, it exports rice to 20 countries including in the EU, the USA, Saudi Arabia and the UAE. In Europe, it is linked to major suppliers (Masterfoods, Soufflet Almentiere, Euricorn and Herba Group) to major supermarket chains such as Sainsbury's and Tesco. In the domestic market it sells rice under the brands Hello Basmati, Gateway of India and Neelkamal. The total quantity sold in 2005-06 was 51,000 MT.

Table IV.1 below gives an account of the costs incurred by the company since the year 2001 under different heads. Premiums payable to farmers 'during' the conversion period and salaries of staff accounted for the largest

Table IV.1 Costs of Organic Production

Year	Cost (Rs.)									
	Survey (Rs)	Staff Salary and other expenses (Rs)	Legal cost (Rs)	Certification cost (Rs)	Premium during conversion period (Rs)	Farmers' training (Rs)	Others (Rs)	Total (Rs)	No. of farmers	Area (Hectare)
2002	0	1,500,000	0	70,000	3,100,000	500,000	0	5,170,000	190	900
2001	480,000	128,000	24,000	131,500	1,550,000	0	25,000	2,338,500	100	600
2003	0	2,000,000	0	100,000	2,075,000	0	0	4,175,000	275	1,250
2004	0	2,000,000	0	310,510	0	0	0	2,310,510	275	1,250
Total	480,000	5,628,000	24,000	612,010	6,725,000	500,000	25,000	13,994,010	275	1,250

Source: Prepared by Katyal (2006); Taken from Alam (2007).

16. *www.Sunstaroverseas.com*
17. *http://sify.com/finance/fullstory.php?id=14022384*

share of the total investment of nearly Rs. 1.4 Crore (Rs.1195/ Hectare). In contrast, note that certification costs are a mere 4.4 per cent of total costs borne, possibly signaling the effectiveness of the ICS.

Table IV.2 Costs of Production in Organic and Conventional Production of Basmati Rice

Activity	Cost Rs/Hectare	
	Organic	Conventional
Bio-fertilisers and manure/chemical fertilisers	2727	3338
Pesticides/fungicides/weedicides	0	1675
Seed	510	480
Farm preparation (tilling)	4200	4200
Puddling	1500	1500
Transplanting	2000	2000
Weeding	3200	1600
Irrigation	1800	1800
Harvesting (cutting, threshing & loading)	3200	3200
Cost of finance	400	900
Total	19537	20693

Source: Prepared by Katyal (2006); Taken from Alam (2007).

Finally, we look at the economic viability of organic farming by comparing profitability between organic and conventional basmati rice production. First we look at production costs in both organic and conventional channels as depicted in Table IV.2. We see here that while organic production results in considerable savings on manure, fertiliser and pesticide expenses the costs of seeds and weeding are substantially higher. On the whole however, organic farming is cheaper than conventional by a measure of 5.6 per cent of the latter. To compare total profitability between the two channels, look at Table IV.3.

Table IV.3 Profitability of Organic and Conventional Basmati Rice Cultivation

	Average yield (Ton/Ha)	Rate (Rs./Ton)	Amount received by farmers (Rs./Ha)	Cost of production (Rs./Ha)	Profits (Rs./Ha)
Organic farming	2.25	15500	34875	19537	15338
Conventional farming	2.62	13000	34125	20693	13432

Source: Prepared by Katyal (2006); Taken from Alam (2007).

Clearly, organic farming is a very profitable venture, as farmers receive both higher prices per unit of output (as measured in tonnes) as well as higher total profits (i.e. from a given plot of land, therefore the yield reduction effect is completely offset by the higher prices in revenue terms). While a hectare of land under conventional basmati production could generate a farmer a profit of Rs. 13,432 the profit under organic farming was Rs. 15,338; representing a 14 per cent higher level of profit.

IV.4.2. UOCB

The role of the UOCB in basmati rice marketing is studied here, drawing largely from two studies: the first conducted by the UOCB for the FAO in Kashipur Block of Udham Singh Nagar district, and the other by Alam and Verma (2006). The origins of the basmati export activities of the UOCB lie in the DASP Programme of the World Bank, upon completion of which the organic basmati activity was taken over by the UOCB. At the outset it should be stated that the UOCB does not operate state-owned farms, nor does it directly enter the marketing sphere; it seeks to act as a catalyst by persuading farmers to adopt organic faming, arranging for certification, providing the necessary training as well as organising buyers and sellers meetings to help marketing activities.

The Board has (as per data available till 2006) undertaken certification for over 969 hectares of land under organic basmati cultivation (this includes 219 hectares certified organic, the rest being under conversion in 2006) across 199 villages, representing 1219 farmer households.

The Board works with farmers through Farmers' Federations (two in each district: Dehradun and Udham Singh Nagar). Farmers desirous of cultivating and marketing organic basmati rice (or indeed other products) may

Table IV.4 Certification Costs in Total Costs of Cultivation*

S. N.	Activity	Amount (Rs./hectare)
1.	External inspection by IMO	194.50
2.	Internal inspection	79.73
3.	Certification charges	274.23
4.	Total cost of cultivation	10479.58
5.	Cost of certification as percentage of cost of cultivation	2.69 %

Source: Alam (2007)

* The certification and production cost figures relate to Kashipur block in Udham Singh Nagar district. However, they are representative of the rest of the area.

Table IV.5 Production Costs of Organic Basmati Rice

Activity	Average Cost Rs. Per Hectare
Ploughing	616.05
Seed	188.10
Nursery preparation	363.10
Transplantation	1806.55
Irrigation	535.73
Manuring	270.25
Pest control	1339.27
Spraying	604.18
Transportation	940.47
Weeding	1086.30
Harvesting	2455.35
Certification	274.23
Total	10479.58

Source: Srivastva (2006); Taken from Alam (2007)

Table IV.6 Profitability of Organic Basmati Cultivation

	Particulars/Ha	Amount
1	Average cost of cultivation	10479.58
2	Production (kilograms)	1837.00
3	Sale value @ Rs. 16.25 per Kg	29851.25
4	Net profit	19371.67

Source: Srivastva (2006); Taken from Alam (2007).

join these Federations after paying a nominal fee of Rs. 50. The member farmers are charged a 2 per cent commission by the Federation for their administrative expenses and also the storage and marketing facilities offered by them. The UOCB often helps forge links between the Federations and buyers, but increasingly it is observed that the Federations have been playing a more proactive and independent role.

UOCB primarily performs a facilitating role by a) motivating farmers to become organic growers; b) provide training and technical inputs; c) arrange for organic certification; c) help in the marketing of organic rice at a premium. The Board has appointed a number of field officers who carry out these functions. Furthermore, the Board has encouraged the formation of farmers' federations. These federations act as links between farmers, the Board and the buyers. The Board does not market the rice directly.

The Board has, on its rolls, a staff of field workers who provide necessary training to farmers in organic farming, and also help in the process of certification. It arranges for certification from international agencies such as ECOCERT, IMO Control and USOCS. To bring down per-capita certification costs the UOCB too has an ICS in place, where UOCB staff conducts internal inspection and only a small number of farms are inspected by external inspectors. According to Alam (2007), the cost of inspection has come down from the erstwhile average of Rs.3000/Man-day to Rs. 350/Man-day. As in the previous case, the cost of certification accounts for a small share in total costs; a tiny 3 per cent. This may further fall as the Uttarakhand State Seed and Organic Production Certification Agency gains recognition as a certifying agency. Table IV.4 breaks down the various components of certification costs from an example of IMO in Kashipur block:

The production of certified organic basmati rice in 2005 was 109.8 tonnes. It received a premium price of Rs. 21000/tonen. The total value of the organically certified basmati rice sold by farmers working with the project was about Rs. 2.3 million in 2005. Table IV.5 gives an indication of various other components of total production costs:

As is clear from the table, harvesting, transplantation and pest control account for the largest shares in the production of basmati rice. We now look at profitability of the organic basmati proposition as depicted in Table IV.6.

The net profit from organic basmati rice cultivation is estimated at Rs. 19, 372 per hectare. Though a comparison in the form of prevailing profits in non-organic plots is not available, we can safely assume this is indeed profitable, as in the previous example the more profitable organic basmati rice yielded a profit of less than Rs.16, 000 per hectare.

One concern with regard to productivity: Alam and Verma (2006) have noted that these profit figures are only for rice. Most farmers typically grow another crop as well in the year (most commonly wheat). It has been observed that after switching to organic farming practices, wheat yields tend to decline drastically, and since there is no ready market for organic wheat willing to pay a premium, the profits on wheat suffer. Thus, the profits emanating from rice exports are dampened by slight losses witnessed in wheat. The decline in profits from wheat was noted to be as much as 35 per cent in Dehradun district.

In the early years, the UOCB helped the Federations link up with Rapunzel, a German buyer of organic basmati rice. In 2004-05, 85 tonnes of organic basmati rice were exported to Rapunzel. Demonstrating market awareness and initiative, by 2005 the Federations felt that they could get a better deal on the produce, they looked for new buyers and found a suitable match in Satnam, a large exporter of basmati rice to whom they sold 175 MT at a price of Rs. 21,000 per tonne. Satnam, through its brand "Kohinoor" is present in 57 countries. Satnam also sells through leading global retail chains such as Tesco, Harrods, Costco, Coles, B.J.'s, Somerfield, Whole Food, Tiger Brands International, etc. In Indian markets too, the Kohinoor brand has a 35 per cent share.[18] Satnam also have taken up the responsibility of procurement and processing of paddy, which earlier was undertaken by the UOCB.

Clearly, in both case studies we see how the niche export markets have been successfully tapped by proactive linkages with key players in the world markets. We now look at the case of a donor-funded NGO intervention which has certified vast areas as organic and seeks to sell all the produce under the organic umbrella.

IV.4.3. INHERE

INHERE is a local NGO active mainly in the Almora district of Uttarakhand since 1982. Having started off as a research organisation focusing on the environment and livelihoods, INHERE has spread over the years its presence and influence across the district, being a grassroots implementing organisation for several donor-funded development projects. In recent years, the organisation has taken up organic farming as a key strategy in its livelihoods enhancement efforts as they believe it is simultaneously most suited to Uttarakhand's poverty alleviation and environmental sustainability concerns.

Under a development assistance program supported by the Indo-Canada Environment Facility, INHERE was offered free certification for organic agriculture in 82 villages in its project area. The certification was provided by the Dutch agency, SKAL International. Other inputs for INHERE's organic programme came from Winrock FTF, the GMED Programme and the SWAF programme. Having completed the certification process for all 82 villages, INHERE runs local processing units for cereals and fruits. This is part of INHERE's larger strategy towards food security, diversified organic agriculture, better land sustainability and higher market prices. We visited both these processing units in Almora district and interacted with producers as well as processing unit officials.

The cereals and spices processing unit has been operational for the last 4 years, and processes and supplies cereals, lentils and spices to buyers based in metros across the country. FabIndia Overseas Ltd., a pioneer of sorts in organic food retailing across its 100 outlets in Indian cities is the major client, singularly accounting for about 55–60 per cent of the unit's sales volumes (estimated at Rs.18–Rs. 20 lakh for 2006–07). Clearly, this is not a large-scale operation but a very promising start for a new entrant in a nascent domestic organic market.

The interested companies approach the processing unit every year with their requirements of particular products. INHERE then, through its network of field workers, assigns targets for particular farers and provides them with the necessary seeds (when introducing a crop for the first time, old seeds are used subsequently) and training on good cultivation practices. Upon delivery of the processed produce, the companies pay the unit at negotiated rates, while the farmers are paid by the unit immediately in cash. It was reported that organic wheat would command a premium of Rs. 5/Kg and rice Rs. 10/Kg at the farmer-level over conventional varieties.

Farmers reported some problems with packaging and storage but hoped that these would be improved in time. There were also very stringent quality standards imposed upon the produce with lower prices for lower quality

18. *http://www.thehindubusinessline.com/bline/iw/2005/09/04/stories/2005090400200900.htm*

produce and outright refusal of bad/rotting produce. However, farmers have been very enthusiastic to sell their produce to the processing unit, and officials claim that supply is not a problem. Although demand has invariably increased over the years, the unit has never faced a supply crunch. The willingness on part of farmers may be understandable as alternative channels for selling organic produce at a premium may not be available.

The fruit processing unit has a turnover of Rs. 30–32 lakh, and processes about 160 quintals of fruit per year. Fruits such as mangoes, peaches, pears, *malta*, lemon, gooseberry and rhododendron (flower) are processed into pickles, jams, concentrates and squashes. Again, FabIndia Overseas Ltd. and Organic Foods India are the major buyers over the last 3 years of the unit's existence. The processed output is also sold in local markets such as in Nainital, in tourist resorts and as raw fruits at the *mandis* when there is no demand to process more quantities. As in the case of cereals and pulses, in the case of fruits too quality considerations are key, especially for pickles.

The farmers received a price premium on organic fruit, ranging between 10–15 per cent. One alarming discovery made over the course of our visit was that a mere 20 per cent of the fruit production of the area was being processed at the unit. The rest 80 per cent, though grown organically was being sold at the local *mandis*, without any premium for its organic status. This stems from the coincidence of inadequate processing capacity of the unit and a surplus production of organic fruit in the area. It is well possible that this may be the case for other crops too, and represents a classic problem associated with *en masse* certification without creating adequate market linkages and demand.

IV.5. Common Threads and Lessons from the Case Studies

The above case studies, though brief, provide an overview of the main forms that organic cultivation could take on in practice, and highlight the strengths and weaknesses associated with each.

In the case of Sunstar Overseas, the simultaneous existence of a strong federation of farmers and the company's strong agri-input distribution network assured gains to both parties. Land for farming was carefully chosen where farmers were not previously using very high amounts of chemicals. Sunstar was able to enter into long term contracts and commit funds over long periods of time, and also offer farmers a complete buyback commitment. Most importantly, Sunstar's position as a big player in world and domestic markets ensured sustainable long term profits to farmers. Such market link played a crucial role in Sunstar's success. On the flipside, the illiteracy of farmers prevents many from participating in the project as paperwork becomes tedious. Also, problems faced due to expensive bio-fertilisers, bio-pesticides and composts have been reported.

In the case of the UOCB, the single most assuring factor for farmers was that it was a government-run initiative, and hence seemed reliable. The UOCB has also encouraged innovation by finding lucrative markets for traditional crops grown by uninitiated farmers. For instance, it was able to identify a Japanese buyer for *manduwa*, an otherwise lowly-priced local millet with a small local market. It has a dedicated team of driven field workers and maintains good relations with farmers. On the other hand, despite their efforts, the intensity of organic cultivation in their project areas remains low. Alam and Verma (2006) show that the percentage of project area land under organic crops has remained stagnant at about 16 per cent over the past few years. The losses faced by farmers in the second crop in the year (usually wheat) alluded to earlier also merit concern and corrective action. It reflects a certain lack of initiative thus far on finding niche markets for organic wheat. Another concern mentioned often was that though there is a dedicated band of field staff, the Board is considerably understaffed, and hence field workers cannot spend much time, which in turn affects quality of certification and training. Again, pest and disease control support was found to be inadequate.

In both these cases however, the high costs of certification have been avoided by putting in place an effective ICS. Certification costs have thus become a small share of total production costs, more so in the case of the UOCB.

The case of INHERE demonstrates both good and bad practices. While on the one hand the organisation has forged very strong market linkages and continues to do so, on the other it is evident that the *en masse* certification exercise, in the absence of large enough organic markets for the local produce, may actually work against farmers as it compels them to sell organic produce at conventional produce prices. Again, the fully subsidised certification has prevented the creation of an ICS, possibly preventing ownership and accountability on part of the local community. On the flipside, the risks associated with organic farming for an individual farmer are mitigated when the entire community adopts conversion; as against the case of the UOCB project areas where the intensity of organic farming remained low. Another advantage of large scale area certification is the subsequent ability to explore organic options in a whole range of crops that grows across the certified area. This is evident from the long list and variety of commodities being produced and marketed from the project area, albeit in small quantities for now.

IV.6. Conclusions

It seems clear that in order to be reaping the benefits promised by organic farming, farmers need hand-holding and federation-type support at least in initial years, to persuade them to practice organic farming in the first instance, and follow good production practices. This support may come from a variety of institutional actors, and the efficacy of such interventions depends not so much upon the institutional character of the organisation *per se*, but more on the role that it plays.

Strong market linkages must be at the forefront of organic farming support programmes. This is especially important when farmers may not have access to any other organic marketing channel for their produce. The decline in yield associated with conversion to organic must be at least matched if not entirely offset by increases in prices, so that the farmer is not any worse off. However, this is not to say that high market premiums for organic products are necessary and/or sufficient conditions for promoting organic farming. Organic farming may still be efficient in the absence of such premia, due to possible reductions on the costs side. The reduced environmental damage and consequential restoration of natural ecosystems can indeed be reasons enough to support organic farming.

Organic farming too calls for a package approach; meaning that when promoting organic farming adequate provisions for bio-pesticides, bio-fertilisers, and compost and also training and extension services pertaining to organic farming. These are currently lacking in the state and more attention needs to be paid for better delivery of these goods and services. Further, research on better techniques of organic farming, and newer niche crops must be encouraged and effectively disseminated amongst interested parties.

Conversion period losses may be particularly discouraging for farmers who recently have used chemical fertilisers and pesticides on their plots. In the absence of additional credit facilities for organic farming, and a generally poor credit market for small farmers, risk aversion may force inertia and discourage organic production. Suitable financial incentives and insurance mechanisms need to be thought of and introduced to encourage organic farming.

In many villages where organic experiments have not met with much success, the shortage of manure and water were commonly mentioned (Mishra, 2006). Manure requirements may often be very high to maintain nitrogen–carbon ratios (as often specified by foreign buyers), calling for very high volumes of cow dung that may not be easily available to small and marginal farmers.

Development interventions must focus on providing much-needed support in the form of extension services. It should be borne in mind that if the purpose of promoting organic farming in a particular area is not to tap into the niche market, certification costs can be rather low. In other cases, these can also be offset due to the higher price premiums. Compensation for certification therefore is not the major constraint, and hence support initiatives should focus more on extension and inputs. The latter would include irrigation as well as a better developed system of providing organic seeds, pesticides and fertiliser. More attention to research on developing better and more varieties of such inputs should also be a thrust area. To conclude, organic farming offers much promise from economic and ecological standpoints, but the exact effects of adoption depend a lot on the policy environment and institutional mechanisms in place. If harnessed well, it may offer a great opportunity to small and marginal farmers to escape poverty, if not it could have debilitating impacts on livelihoods and incomes of equally numerous and vulnerable households.

Annex V

Medicinal, Aromatic and Culinary Herbs

V.1. Introduction

The altitude, agro-climatic and topographical conditions of Uttarakhand while encumbering large scale cultivation for traditional cereal crops (unlike the rest of India) bestows upon the state clear advantages in the cultivation of many non-traditional crops, chief among these being the umbrella category of herbal plants (including medicinal, aromatic and culinary herbs). In recent years this sector has received much interest and attention for its potential as a lucrative income-diversifying and income-supplementing option for Uttaranchal's rural and forest-dependant communities. Globally rising demand for herbal products and raw material also project the possibilities of developing a strong export-oriented herbs production and processing industry; potentially contributing to the state's growth trajectory and development goals in the years to come. This particular annex attempts a holistic overview of the issues in the, cultivation and processing of herbal plants in the state and of government and private efforts to spur innovation and growth. Based on qualitative interviews with numerous stakeholders in the field, as well as drawing from recent research studies on the value chains of herbal crops, we seek to identify the priority areas that public policy must address, as well as highlight certain successful case studies.

Natural forests and rangelands account for about 70 per cent of Uttarakhand's land. Local communities, largely living in high reaches of the Kumaon and Garhwal hills are very intimately dependent upon forests and forest products for their subsistence and livelihoods needs. The people of the state have a long history of dependence on the large varieties of medicinal herbs abundance, drawing from ancient *Ayurvedic* traditions. The local knowledge on the use and application of these is greatly augmented by cutting-edge research on these undertaken increasingly by corporations in response to rising demands in domestic and international markets; which in turn necessitates sustainable supply of a large basket of ingredient herbs in herbal medicinal value chains. Blessed with a mélange of sub-tropical, temperate and alpine ecosystems, Uttarakhand has the unique advantage of being home to a large number and variety of herbal plants.

At the all-India level, the main centres of bulk trading in medicinal and aromatic plants are Delhi, Amritsar, Raipur, Chennai, Kolkata and Mumbai (Subrat *et al.*, 2002). In Uttarakhand, larger volumes of trade take place at Dehradun and Tanakpur, and to a lesser degree at markets in Ramnagar and Haridwar. A new herbal *mandi*, proposed at Haldwani is expected to become operational soon. These are designated centres, run and operated by the Forest Department where notified herbal produce (i.e. legally cultivated, or legally collected, up to a specified extent from forests; to prevent excessive collection) may be traded through a system of auctions where commission agents and traders compete to determine a competitive and remunerative price. The sources of medicinal plants traded in India according to a senior Ministry of Health and Family Welfare, GoI official (Sajwan, 2007), are given in Table V.1:

Table V.1 **Sources of Medicinal Plants Traded in India**

Source	Percentage of total medicinal plants
Wild	61
Wild/cultivated	25
Cultivated	10
Exotic	4

Source: Sajwan (2007).

Although no precise statistics of such kind are available at the state level, it is worth remembering that, given the wide prevalence and use of medicinal plants for own use by Uttarakhand's local communities, the importance of people's reliance on forests (here only for medicinal plants) can only be understated if we draw conclusions from data on traded volumes alone.

V.2. Global and National Demand in Derbal Industry

The herbs industry is often thought of as a lucrative income diversifying option, owing to an increasing domestic and international demand. First, we attempt to gauge the domestic demand for herbal plants, excluding own use by cultivators and collectors alike (which on the whole may be rather small in volume). An estimate of the national demand for various herbal products in the years 2000-01 and 2004-05 and annual growth rate computed from there is given in Table V.2.

Ranging between 1.8 per cent to as high as 30 per cent, the annual growth of demand witnessed for herbs over the years is astounding and at first glance, indeed encouraging for suppliers. Specifically for medicinal plants, it is estimated that about 70 per cent of India uses traditional medicine, especially the rural poor, tribal and other forest-dependant populations (who may typically be using local herbs in crude forms, and not prepared formulations). The rising demand is also reflected in the sales volumes of major herbal pharmaceutical companies. Table V.3 shows the most recent available figures of sales volumes of major players in this sector comparing the years 1993-94 and 1998-99. Although somewhat outdated, Table V.3 gives an indication of the spurt in sales (including exports, however since these are a small fraction of the total sales the figures can be treated as representative of domestic sales volumes).

Based on an extensive survey of agents and manufacturers, Subrat *et al.* (2002) calculate an annual compound growth rate in sales of 20 per cent. They go on to summarise estimations of national demand for various species of herbal products provided by different agencies, and conclude that there are vast variations in figures estimated by different agencies and argue for the need for a more systematic survey at the national level.

The export markets also hold much promise. Exporting 33,900 tonnes of herbal raw material, India is the third largest exporter of herbs in volume terms in the world, but lags behind considerably in value terms (the volume represents US$ 57 million). To illustrate, consider the case of Singapore whole sales of 9,600 tonnes represents a value of US$ 567 million. This indicates the extent to which specialised species as well as some value addition at initial stages can help extract maximum

Table V.2 Domestic Demand Estimation for Key Herbs

Biological name	Common name	National demand (MT)		Annual growth rate %
		2000-01	2004-05	
Aconitum heterophyllum	*Atis*	270.1	448.4	18.4
Swertia Chirayata	*Chirata*	965.2	1284.7	10
Curciligo orchiodes	*Jatamansi*	674.9	866.8	8.7
Gloriosa Superba	*Kalihari*	65.4	100.5	15.4
Saussurea superba	*Kuth*	1414.1	1826.3	8.9
Picrorhiza kurrooa	Kutki	220.3	317	12.9
Rauvolfia serpentina	*Sarpagandha*	423.6	588.7	11.6
Emblica officinalis	*Aaonla*	22729.5	41782.8	22.5
Withania somnifera	*Ashvagandha*	7028.7	9127.5	9.1
Phyllanthus amarus	*Bhuin Aaonla*	2212.6	2985.3	10.5
Bacopa monnieri	*Brahmi*	3822.5	6621.8	20.1
Berberis aristata	*Daru Haldi*	1187.3	1829.7	15.5
Tinospora cordifolia	*Giloe*	2258.3	2932.6	9.1
Andrograpis paniculata	*Kalmegh*	2005	2197.3	3.1
Solanum nigrum	*Makoi*	2077.9	2192.2	1.8
Glycyrrhiza glabra	*Mulhati*	873.4	1359.8	15.9
Coleus barbatus	*Pattharchoor*	37.8	60.8	17.2
Piper longum	*Pipli*	3992.5	6280.4	16.3
Cassia angustifolia	*Sanay*	6462.5	11677.3	21.8
Asparagus recemosus	*Shatavari*	10942.7	16658.5	15.1
Ocimum sanctum	*Tulsi*	3296.8	5402.9	17.9
Aconitum ferox	*Vatsanabh*	322.3	3426.8	30
Saraca asoka	*Ashok*	7051.3	10724.2	15
Aegle marmelos	*Bel*	5381.2	7084.5	9.6
Santalum album	*Chandan*	635.2	1073.1	19.1
Commiphora wightii	*Guggul*	1505	2548.9	19.2

Source: Chauhan (2006).

Table V.3 Sales Figures for Key Herbal-based Pharmaceutical Companies

Company	Sales (Rs. Million) 1993-94	Sales (Rs. Million) 1998-99
Dabur India Ltd.	2610	3220
Zandu Pharmaceuticals	390	1200
Himalayan Drug Co.	310	1400
Baidyanath Ayurved Bhavan	-	1360
Ajanta Pharmaceuticals	170	200

Source: ITCOT (1999). Table reproduced from Subrat *et al.* (2003).

Table V.4 **The Top Ten Exporters and Importers of Herbal Producers in the World**

Exporters				Importers			
Rank	Country	Volume (MT)	Value (US$ Mn)	Rank	Country	Volume (MT)	Value (US$ Mn)
1	PRC	147000	282	1	Hong Kong, China	67000	291
2	Hong Kong, China	63150	229	2	Japan	51350	136
3	India	33900	57	3	USA	49600	136
4	Germany	15100	70	4	Germany	45350	110
5	USA	13500	116	5	Korea, Rep. of	32250	52
6	Mexico	13000	11	6	France	21350	52
7	Egypt	11750	15	7	PRC	13650	41
8	Chile	11600	28	8	Italy	11700	43
9	Bulgaria	11050	15	9	Pakistan	11050	11
10	Singapore	9600	567	10	Spain	9100	28
11	Morocco	8000	13	11	UK	7650	27
12	Pakistan	7800	5	12	Singapore	6300	51

Source: Vogtmann (2004) Reproduced from Sharma (2007)

benefits even when volumes may be low. Volumes and value of trade of the world's largest exporters and importers of herbal produce are given in Table V.4.

Overall, the international market for medicinal-plant based products is estimated at US$ 60 billion and is growing at the rate of 7 per cent per annum (Ramakrishnappa, 2002; Planning Commission, 2000; CTA, 2002). More specifically, in the US market, herbal products worth US$ 4 billion were sold as long back as in 1996, which figure was expected to double by the year 2000. The newly emerging niche neutraceutical market (nutritionally or medicinally enhanced foods with health benefits) is worth as much as US$ 80–250 billion in the US and EU each, and about US$ 1.5 billion in Japan. This growing market represents a window of opportunity for herbal exporters (though herbs here would be ingredients in and not the mainstay of such diet supplements).

In order to tap these export markets, however, one has to go through a number of channels till the product may reach the end user. A typical export supply chain for herbal products is depicted in the diagram in Figure V.1.

Apart from the procedures involved in export of all commodities of agricultural original, there are a few additional factors that constrain access to world markets, and substantial and sustainable earnings from them. First, most of the herbal products, especially medicinal herbs exported by India, as of now are done so in the crude/raw state. This is because processed medicines/formulations are allowed to be exported only if the particular formulation/medicine is registered in the destination country. Generating technical and clinical evidence to ascertain the medicinal effects and permissibility (for medicinal use) is an expensive affair and cannot be profitably undertaken in most cases. Most herbs therefore end up being sold in raw state, to be used subsequently as diet supplements.

Conservation efforts, leading to drawing up the "negative" list of 29 species of plants banned for exports also in the short-run curtails exports options. However this may be revised in time as it was in 1998, when the negative listed species reduced from 53 to 29 species, and restrictions on exports of formulations of these products were lifted. The Convention on International Trade in Endangered Species (2002) also has an explicit list of species for which international trade is banned, and typically these are traded illegally, albeit in small volumes and at very high prices.

Figure V.1 **Export Supply Chain of Herbal Plants**

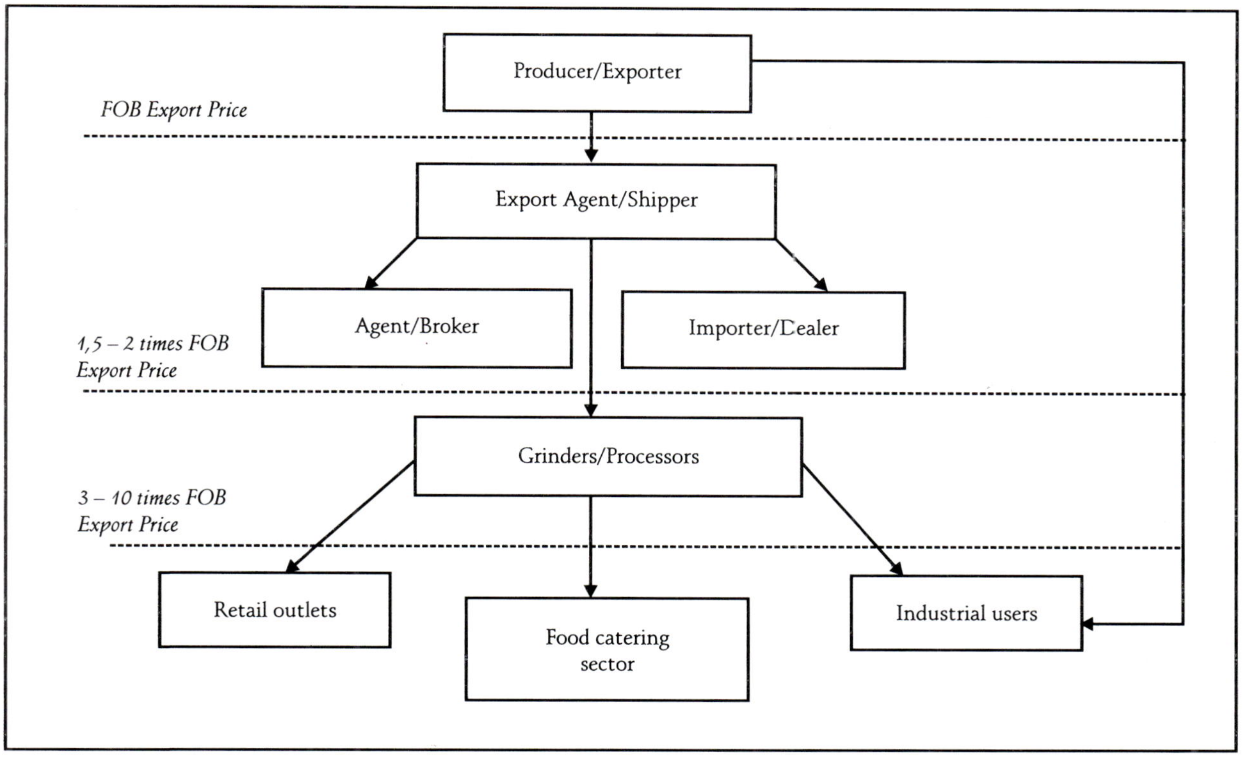

Source: Sharma, R. (2007)

V.3. Institutional Channels of Production and Marketing

In the context of large volumes of medicinal and herbal plants growing in the wild (and with a considerable share of plants that are both wild as well as cultivated) the production and marketing arrangements for such crops exist in a unique setting, with considerations of environmental damage, unsustainable collection and local extinction featuring prominently, very distinct from issues in production of traditional crops. Efforts to strengthen the cooperative movement as early as 1949 led to the creation of Bheshaj Sanghs (medicinal plants growers' unions), who along with private contractors engaged in the business of medicinal plants collection. Although collection increased and incomes were generated, one also witnessed collection far in excess of sustainable levels, disregard of established rules and norms and in some cases also the extinction of certain species of medicinal herbs. Further, the government acknowledged the vast losses in tax revenue due to large volumes of illegal trade that were flourishing in the state. In the year 2003, the Government of Uttarakhand in response to the widely felt need for a better regulated and managed system of production and marketing of medicinal and other herbal plants entrusted the Uttarakhand Forest Development Corporation (UAFDC) with this responsibility.

The UAFDC has developed a multi-tiered collection/cultivation and marketing model for the state, whereby local bodies including the Bheshaj Sanghs, Van Panchayats (loosely, forest protection committees) and the Kumaon Mandal Vikas Nigam (KMVN) are at the forefront of bringing together collectors and cultivators for organisation and management on more sustainable lines. Different areas are earmarked for the jurisdiction for each of these bodies who are then responsible for collecting the farmers' produce/collection from the area, consolidating the produce and then bringing it to the *mandis* for sale through commission agents to traders. Farmers may individually also turn up at the *mandi* for sale though in practice this may not be very feasible. However, in order to collect from the wild, as well as cultivate plants that also may grow in the wild, one still needs permits issued by the forest department and made available through one of these agencies. The Herbal *mandis*, run and operated by the Forest Department conduct auctions on two specific dates of the month. The

Table V.5 Collection and Marketing Report: Combined Uttarakhand Herbal *Mandis*

Year	Quantity (in Quintals)	Revenue (INR)	Collecting agencies
2003-04	2,780	5,967,379	Forest Development Committee (FDC)
2004-05	14,634	62,726,040	FDC, Bheshaj Sangh, KMVN
2005-06	22,111	93,415,133	FDC, Bheshaj Sangh, KMVN
2006-07*	3,186	32,086,180	FDC, Bheshaj Sangh, KMVN

Source: Ratanpal and Singh (2007)
* 2007 figures are only until February 2007

auctions are well-publicised in local newspapers beforehand to bring it to the notice of interested parties. The rate of commission to be charged to the collecting agency by the commission is fixed at 1 per cent of the total sales price (to the collecting agency). The payment for this is made by cheque or bank draft and the clearance is given immediately by the UAFDC. The quantities of herbal products collected through these channels, and the value they represent are summarised, for the last 4 years since inception of the new institutional arrangements in Table V.5.

Having declared itself a Herbal State, and having adopted institutional mechanisms to the effect of better management of the collection and cultivation of medicinal and other herbal plants (discussed above), the Government of Uttarakhand displays a keen interest and concern for the protection and promotion of wild and cultivated herbal flora of the state, and a belief in its potential benefits on income and employment generation. A number of government policies have therefore been directed toward benefiting this sector.

V.4. The Medicinal and Aromatic Plants Supply Chain

The prescribed channels of medicinal herbs trade within the state, as envisaged and facilitated by the state government, linking collectors and cultivators to buyers through unions, cooperatives, parastatal agencies and the establishment of dedicated herbal *mandis* monitored by the Forest Department has been discussed in the previous section. The fact that the herbal plants' value chain is different from that for a typical agricultural product due to the large volumes of produce collected from the wild and the predominance of informal, and often illegal channels of marketing pose particular challenges in using very precise value chain analysis to understand the issues and constraints in production. These issues are addressed more thoroughly in the following section, and for now we focus attention on understanding the broader channels of production and sale of medicinal and aromatic plants in Uttarakhand.

It is estimated that a mere 10 per cent of the total volume of medicinal plants are cultivated, and individual collections from the wild account for the rest 90 per cent (Sharma, 2007). Although these estimates cannot be treated as very precise due to large unaccounted illegal volumes, it gives an indication of the larger role of the Forest Department as compared to the agricultural *mandis*. The Department identifies particular areas from which herbs may be collected and fixes ceiling levels for procurement of particular species. Local cooperatives (mentioned previously) are given fixed quotas for the same. Thereafter, the cooperatives employ contractors who in turn organise local villagers to identify and procure fixed volumes of the particular species. The collected material is sold by the cooperatives to traders and processing industry.

Table V.6 Key Actors in the Wholesale Trade of Medicinal Herbs: Delhi and Mumbai

Players	Delhi	Mumbai
Commission agent	100	100
Exporter	15 – 20	> 50
Wholesale dealer	0 – 35	33 – 35
Trader	30 – 35	30 – 35
Retailer	> 100	≃ 100

Source: Subrat *et al.* (2002).

Traders may either themselves sell consolidated volumes of herbal material to processing pharmaceutical companies, or sell the material in raw form to interested buyers. The profusion of small-scale local practitioners of *Ayurveda* in Uttarakhand and outside, along with an export market which is also interested in raw forms rather than formulations of specific herbs accounts for the latter. The largest wholesale markets for herbal produce in the country are located at Delhi and Mumbai. Subrat *et al.* (2002) attempt to provide a rough estimate (based on their own survey) of the numbers of wholesale traders and other agents involved in medicinal herbs trade at these two centres. The estimates are given in Table V.6.:

Figure V.2 Structure of the Herbal Plants Supply Chain

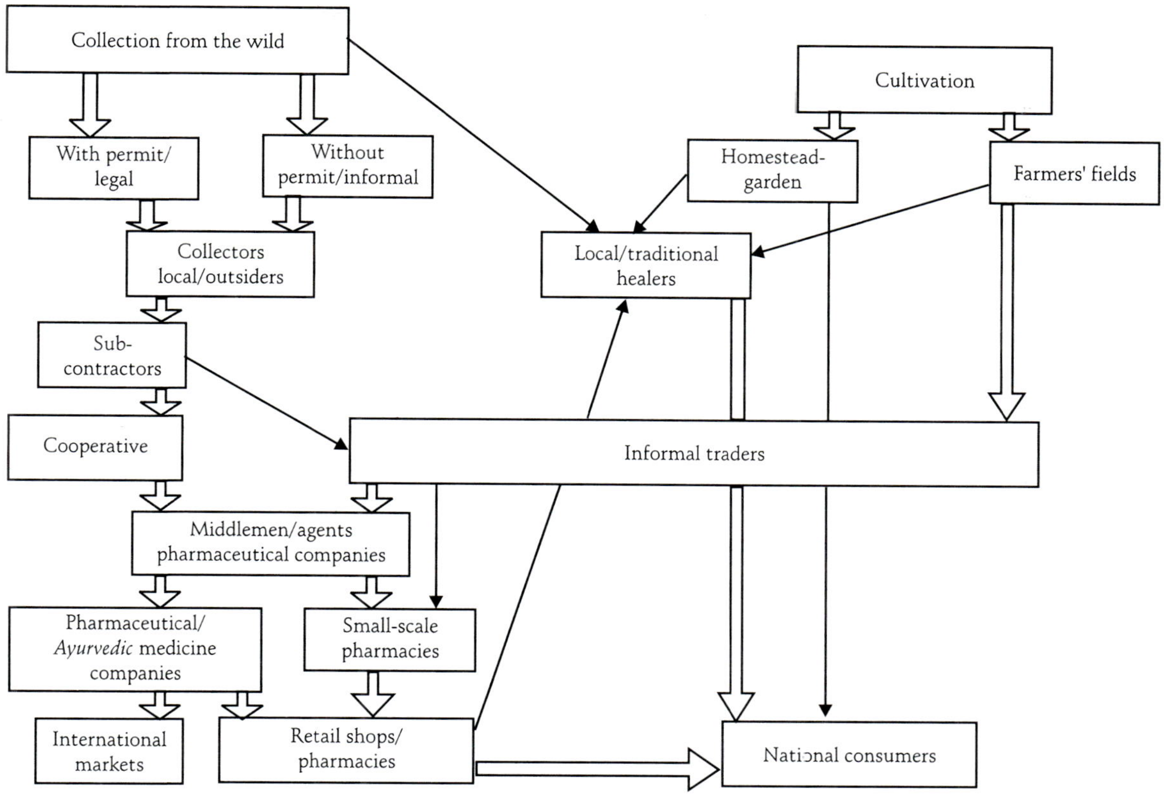

Source: KIT (2003)

On the processing front, one estimate suggests that there are about 7,800 drugs manufacturing units operating in India (Planning Commission, 2000). Agents of the larger companies regularly procure raw material through a network of agents and listed suppliers. It is further estimated that processing companies use 400 species of herbal plants, of which 35 are important (ibid.). Figure V.2 depicts the structure and channels of the herbal industry supply chain, clearly indicating various types of private and institutional players across the spectrum ranging from individual cultivators and collectors on the one end to retail end buyers of raw and processed herbs in both domestic and international markets.

Much recent literature (for e.g. KIT, 2003; Subrat *et al.*, 2002) as well as interactions with small cultivators, processors and government and NGO officials reveal that there is a wide acknowledgement of many supply chain bottlenecks and inefficiencies. First, it seems that collectors are often not aware of the prices and demand prospects for particular herbs, and are consequently cheated and misled by private traders. This is true at all levels of the value chain and invokes the pressing need for public efforts to provide accurate, detailed and timely information on prices of herbal products, to provide indicative benchmarks and strengthen agents' bargaining positions. It is noteworthy that unlike with traditional crops, the absence of state-determined Minimum Support Prices for herbal crops means that there is absolute no benchmark for farmers to weigh their options against. The largely informal nature of the trade further impedes much corrective action.

The overlapping of control of the APMC and the Forest Department (herbal) *mandis* has also been highlighted as a problem (Singh, 2007). Certain medicinal and aromatic herbs have wider applicability to a whole range of other uses and have been categorised for treatement under the APMC Act, which impedes their trade and value for medicinal (or oil extraction purposes) by imposing higher levels of tax (APMC *mandi* taxes and commission rates are typically higher than in the herbal *mandis*).

An empirical study conducted by KIT (2003) documents problems associated with obtaining permits for cultivation and marketing of wild herbs (i.e. those that usually grow in the wild but could also be cultivated). They highlight ambiguity in the procedure and eligibility criteria for obtaining licences, also mentioning that sometimes farmers were clueless about whether they had to obtain licences from the Forest Department or the designated cooperatives.

Despite meeting eligibility criteria, many farmers reported that they had not been given the promised licenses. Our own interactions with farmers also revealed that even when farmers had valid licences, they were often interrogated and harassed by Forest Department officials who would allege that the material had in fact been collected from the wild and demanded bribes, or confiscated the material. Farmers in Almora district told us that they would rather surreptitiously transport the herbal material from the fields to the markets or processing units, in small amounts in cloth bags, rather than *en masse*, to avoid interception and harassment by Forest Department officials. This implies repeated travel to sell produce, which is particularly costly for non-perishable herbal products such as tree barks, which can be amassed and then sold all at once. A discouraging environment such as this must definitely deter farmers from cultivating herbs and warrants immediate intervention.

Even the institutional mechanism established for selling wild herbs through cooperatives to generate scale and volume economies is fraught with problems. The cooperatives may not be able to access up-to-date market information, which increases market uncertainty and farmers' risk. Farmers have very little or no information on how to directly access market outlets and efforts to enhance farmers' direct market access and to improve their viability positions have been abysmal. Subrat *et al.* (2002) argue for catalyst programmes to better organise farmers, provide micro-credit services and organise cultivation on modern management lines with strong market linkages and buy-back arrangements with procuring companies and other buyers. They also mention that more value-addition and processing activities at the local level will enhance farmers' options and strengthen their economic position.

The creation of herbal *mandis* has visibly improved storage facilities at the state level, however more needs to be done at the local levels, where it is reported (ibid.) that much produce especially in the monsoon season goes waste due to damage from moisture; harming both volumes as well as quality of the material.

On the inputs side, studies have reported problems with planting material and poor extension services. The KIT study (2003) reports that often adequate amounts of planting material for certain herbs are not made available in time, which curtails large-scale cultivation. Even the quality of planting material provided is often of bad quality. The study points out one particular case in Chamoli district in which seeds provided for *Kuth* by the cooperative were of poor quality and even failed to germinate!

The framework for herbal plants extension services also appears to be extremely weak. Farmers we spoke with in Almora district reported absolutely no support received from government extension agents, and that technical support came in the form of guidance and direction provided by INHERE, a local NGO. Traditional cultivation of medicinal plants in the state draws largely from deep local knowledge developed and maintained through centuries-old traditions of *Ayurvedic* practice. Subrat *et al.* (2002) point to the alarming involvement of untrained labour in the collection activities which is resulting in considerable damage to the plants as well as the environment at large. Intervention efforts must seek to document the existing local knowledge; and extension services must seek to harness rather than provide an alternative to the cumulative knowledge of local communities.

V.5. Sustainability of Medicinal Plants Promotion: Cultivation *versus* Collection

Unabated extraction of herbs (especially medicinal) leading species to become endangered, and in some cases also extinct across the country is well documented for over the last two decades (Jain 1987, FRLHT 1997, Subrat *et al.*, 2002). Extinction as documented in the case of *Ratanjot* (*Arnemia Benthamii*) and *Bachnag* (*Aconitum Violaceum*) in the Northwest Himalayas, including in Uttarakhand, often has dire consequences as particular species are endemic to specific ecosystems and cannot be cultivated in other ecosystems (Subrat *et al.,* 2002).

Recognising the need to conserve certain species, the Government of India has released a "Negative List" of 29 species for which trade (particularly exports in crude form) is banned. It is noteworthy that of these, 13 species are found in Uttarakhand (Singh, 2007). Although the dominance of illegal trade prevents accurate estimation of existing demand-side pressures on particular species of medicinal and aromatic plants, the Planning Commission (2000) has attempted an exercise to this end. Table V.7 captures the estimated quantity required of particular species found in Uttarakhand, and the period of shortage of supply in years.

Table V.7 **Key Medicinal Plants in Short Supply**

Botanical name	Common name (Hindi)	Quantity required (tonnes/ annum)	Period of shortage of supply (years)
Aconitum heterophyllum	*Atis*	0.55	15
Curciligo orchiodes	*Kali musli*	2.25	4
Hedychium spicatum	*Kapur kachri*	0.225	5
Gentiana kurroo	*Karvi*	0.22	4
Curciligo orchiodes	*Jatamansi*	0.66	5
Picorrhiza kurroa	*Kutaki / Metha*	1.55	5
Plumbago zeylanica	*Chitrak*	3.5	5
Rubia cordifolia	*Manjishtha*	1.15	4
Swertia chirata	*Chirata*	2.5	7

Source: Extracted from Subrat *et al.* (2002), data from Planning Commission (2000).

In the case of specific herbal plants another dimension worth remembering is that these may typically have long gestation periods. While for the collector this implies that collection from wild should take place at sustainable rates and much restrain be exercised in the face of tempting current prices; for the cultivators (where possible) these represent high-value high-risk options.

What then, is the possible scope of cultivation of these crops? What role could they potentially play in the state's development for the aims of agricultural diversification and poverty alleviation? Would the unchecked collection from the wild continue to render cultivation an unattractive or even infeasible option? Further, given a high degree of unsustainable collection, should public policy encourage cultivation of herbs only because it has failed to effectively organise collection, or is there indeed a sound economic basis for advocating such agricultural diversification?

These questions merit critical thought and much more research than is currently available to provide very substantive answers. What is indicatively clear is that for many herbs that require high investments, especially of time, there probably may not be too much of a welfare gain to farmers by switching to cultivation of these from traditional, or more lucrative modern crops. Further, given the current administrative capacity and (rather safely) assuming that connivance of forest officials with collectors and buyers may not be drastically checked in the years to come, one could reasonably expect collection from the wild to remain an important source of supply. Many local communities depend entirely on sales of herbs collected from the wild for their cash economy, and this must be borne in mind when encouraging cultivation as an alternative. Subrat *et al.* (2003) argue that a vehement policy of cultivation could in fact deprive a lot of marginalised, forest-dependant (often effectively landless) communities of livelihoods; and that such an approach would not ensure successful conservation if it antagonises primary stakeholders. They further suggest that small herbal processors are more likely to prefer collection to cultivation due to the lower costs (cultivation costs are nil) and risk involved with this, especially since the volumes they wish to procure may be small. On the other hand, while big processors may find it profitable to switch to cultivation (due to economies of scale), they may typically enter into agreements with a few large farmers, than a scatter of small ones across hardy, mountainous terrain.

Cultivation, in general, may be fraught with a number of other problems as well. At the all-India level, it is estimated that of the total near 500 species used, a mere 70 may be used by a typical herbal pharmaceutical in quantities in excess of 10 tonne (dry). As the remaining species may not be feasible to cultivate they are collected, in the most part, from the wild.

V.6. Paving the Path for Sustainable Herbs Cultivation

The prospects of and issues in herbal crops cultivation identified above must be borne in mind while framing suitable policy and programme interventions for development in the state. It is clear that while there may be very encouraging market signals for the herbal market as a whole, these may not often translate entirely into gains for the poor, especially when we consider the opportunity costs of plots of land which could be used to cultivate other crops. While on the one hand, the efficacy of a cultivation-focused policy is suspect in light of cheap alternatives (collection from the wild), the sustainability of collection practices merits serious thought from a livelihoods as well as ecological standpoint.

Designing and implementing an effective programme would call for a more disaggregated view of the sector, unbundling the mass of "herbal plants" down to specifics, and calculating market viability, sustainability and income-generating potential separately for particular types of herbal produce and extract the state may have a comparative advantage in. This section attempts to put together brief case studies of successful ventures in the state, giving an indication of what works on the ground and under what conditions, how niches are identified, built upon and sustained and finally, what are the lessons these may teach us for emulation. Specifically, we look at the efforts of Flex Foods Pvt. Ltd., a private company near Dehradun engaged in virtual contract farming and export of high-end culinary herbs to European markets, and a Medicinal Plants Processing Unit set up in Almora district, supported by a local NGO, "INHERE" as case studies.

The organisation of collection and cultivation of medicinal and other herbs has been the subject of a keenly contested debate. This involves issues pertinent to people's livelihoods, community interests, environmental sustainability market forces and state policy. Having discussed previously, some of the issues in cultivation and collection of herbs, we now look specifically at the tensions between collection and *in situ* cultivation, possible means of resolution and within this, particularly the role of federations and unions.

Environmental sustainability is a critical consideration for collection from the wild. It has been observed that the lure of quick money often encourages communities to conduct collection at unsustainable rates. A brief look into the profile of the main groups that collect herbs throws light on their peculiar circumstances. According to a study by Virdi (2004) collectors are typically engaged in large amounts of casual labour, have small land holdings, have few or no family members who have jobs outside family-run agriculture, and have a smaller share of family earnings from natural resource-based occupations. It was also noted that members of Scheduled Tribe communities and women were much more likely to be involved in collection of herbs from forests and common lands. In essence, collection is really practiced by smaller and poorer rural populations/groups and is presently considered to be an inferior proposition as compared to most other economic activities.

Given the economic deprivation of collectors and in light of mounting burden on endangered herbal species in particular and wild flora in general, *in situ* cultivation and sustainable collection have often, in policy discussions and practice, been suggested as the two important strategies to address concerns. However, this may be fraught with a number of technical problems on the ground which merit acknowledgement. As has been discussed previously, given the long gestation periods of many herbal plant cycles (especially where entire plants or the roots may be of use), cultivation will always be a very expensive option and the temptation to collect from the wild instead would be high. Cost-sharing between different members of a community in *in-situ* cultivation is also a potential area of conflict, where heightened tensions and disagreements, in the absence of effective governance mechanisms could sabotage cultivation efforts. In such a situation one is constantly faced with the problem of free-riders on investments made for *in situ* cultivation. Likewise, disputes on the ways in which to share the benefits of cultivation, and any improvements undertaken subsequently could also disrupt the necessary coordination for successful *in situ* cultivation.

Given the rampant out-migration from many Uttarakhand villages, it has also been pointed out that there may be a disincentive to collect at sustainable rates spread out over time, and many households may choose to collect (often dig up) herbs when family labour is abundant. To not do so from common lands could in fact be problematic as households face the risk of losing out to other community members who may compete with them and collect (dig out) very large quantities of herbs at a point. This is a classic example of how the first mover advantage could dramatically affect a particular household's fortunes. It has also been pointed out that many a time herb traders give the collecting and cultivating families large advances of money for inputs as well as personal uses, and in turn engage them in tacit contracts to sell them specific quantities of particular herbs. In turn, this has increased the pressure on collectors to extract as much herbal material as possible to pay back loans and continue to be eligible for further loans in the future. Clearly such a situation flourishes due to an ineffective credit market and the poor credit-worthiness of asset-poor rural communities. Note, that these issues are more relevant to discussions on collection of rarer/endangered herbs. Those in abundance and which regenerate quickly do not warrant as much a degree of concern from the sustainability point of view.

In order to address these issues, a few fundamental issues need to be considered. Credit market failures, alluded to earlier are an important area of concern. However, simply increasing capital outlays and loan disbursements may not bring about large improvements in the herbs collection and cultivation processes. First, the scale at which these crops are grown in the state is miniscule, and therefore it is very likely that large credit outlays would be channeled to other uses. Availability of good quality planting material was mentioned as a factor far more important for sustainable herbal plants cultivation than credit alone.[19] The possible role of the government in promoting cultivation is the matter of a contested debate. While some believe that subsidies may be welcome and warranted in the initial stages of cultivation of herbs; many also feel that such a subsidy-culture could distort markets and keep the farmer from aligning with market conditions.

Any intervention; of the collection or cultivation kind must actively involve local communities as it is their local environments and livelihoods that are immediately and directly impacted. This is all the more important as local communities are repositories of long-standing and valuable local knowledge and values concerning the use of many herbs. Traditional knowledge and practices of the local communities should be documented and built into larger programmes. The role of local federations, unions and NGOs is critical in this regard. Created in 1977, cooperatives called "Bhaisaj Sanghs" are operational in each district of the state with the purpose of developing medicinal plants and organising collection and cultivation activities. Their efforts are complemented by a number of local NGOs who are involved in this area of work for environmental, medical and livelihoods motives. Synergy between various institutional actors and cultivators and collectors is crucial for the success of herbal cultivation, marketing and processing in the state of Uttarakhand.

V.7. Case Studies

V.7.1. The Case of Sanjeevani Ayurvedshala

INHERE is a local grassroots-level NGO, active in the Almora district of the Kumaon division since 1982. Concerned with agricultural diversification, food security, environmental sustainability and livelihoods protection, the NGO has identified medicinal plants cultivation as an important component of its rural development initiatives. To promote this, INHERE has set up a medicinal plants processing unit, "Sanjeevani Ayurvedshala" in Masi, Almora. The Sanjeevani Ayurvedshala sells processed medicines and food supplements to three main buyers. These are Vedicare (a French company selling ayurvedic health and beauty treatments), FabIndia Overseas and Winner International. Of the total value of herbs processed at the unit, Vedicare accounts for 65 per cent, the rest being equally divided between FabIndia and Winner. The unit uses the field workers' network of INHERE to plan production to organise farmers and give them production targets on based on a 3-6 months market plan, based on the demands of the buyers. Internal targets between farmers are decided by INHERE field staff based on farmers' land holdings and assessed capacity. For the last year (2006), the unit processed 8 tonnes of herbal raw material. The breakup across different medicinal plants and the value this represents is given in Table V.8.

19. Also note that the class of cultivators is generally different from the collectors; the latter being typically poorer and in greater need of credit.

Table V.8 **Quantities Processed and Procurement Prices of Medicinal Herbs: Sanjeevani Ayurvedshala, Almora District**

Raw herbs	Scientific name	Quantity procured (kgs)	Rate offered (Rs./kg)	Total value (Rs.)
Shatavari	*Asparagus Racemossus*	1200	12	14400
Purnanva	*Boerhavia Diffusa*	35	30	1050
Brahmi	*Cantella Asiatica*	600	20	12000
Aaonla	*Emblica Officinalis*	2500	8	20000
Tulsi	*Ocimum Sanctum*	30	25	750
Bhuin aaonla	*Phyllanthus Urinaria*	300	15	4500
Giloe	*Tinospora Cordifolia*	800	10	8000
Baheda	*Terminalia Bellirica*	1000	12	12000
Harad	*Terminalia Chebulia*	1500	10	15000
Ashwagandha	*Withania Somnifera*	125	60	7500
Goksha	*Tribulus Terrestris*	30	30	900
Total		8120		

Source: Figures provided by officials of Sanjeevani Ayurvedshala from accounts.

Clearly, the volumes as well as value represented by this unit are not very large and are unlikely to have much of an impact even in the local area. Our interactions with medicinal plant farmers provided further insights. Farmers mentioned that INHERE officials were prompt and efficient in providing training and informing them on their targets well in advance. The NGO has arranged for certification of the cultivated produce for the farmers through the HRDI, Gopeshwar, and have also registered with the Forest Department. The certified farmers are about 900 in number. These are supervised and trained (when needed) by 20 field officers, each responsible for a cluster of about 5 villages.

Farmers mentioned that medicinal plants cultivation had received a fillip since the processing unit came up, although the volumes traded are still very small. The chief benefit identified was that the processing unit provided them with close market access to sell their herbal produce as the herbal *mandis* are very far. The medicinal plants were considered an important supplementary source of income, and not the mainstay. The farmers who were cultivating medicinal herbs were doing so in small tracts of their total land, and continued to grow their traditional crops as well. Impacts of medicinal plants cultivation, if any, on subsistence needs were reported to be minimal. Most cultivation patterns remained unchanged. The farmers also reported that very small numbers of farmers in every village had chosen to cultivate herbs, and these too were predominantly those farmers who traditionally have been *Ayurvedic* practitioners. As the processing demands on the unit are rising annually, so is the level of farmers' participation. Many prefer to sell their herbal produce to the unit as it saves them transport costs, and also as they feel that the prices offered are fair and remunerative. Despite certification, farmers pointed out the high-handed behaviour of certain Forest Department officials, who would intercept their movements (when travelling with larger quantities of herbal matter) and harass them. The farmers reported to be earning a supplementary annual income between Rs. 4,000–Rs. 8,000 from herbal sales (only accruing to the unit, and possibly some more from sales at the *mandi*, and to private agents etc.).

V.7.2. Flex Foods Ltd.

Flex Foods Ltd. is one of the largest exporters of processed agricultural products in Uttarakhand. Starting primarily as a processed mushrooms exporter in 1992, Flex has over the years, added to its product portfolio and the business currently includes freeze dried as well as canned mushrooms, culinary herbs, spices, fruits and vegetables. The business has an estimated turnover of Rs.1.4 billion (Culinary herbs account for an estimated 60 per cent) and is constantly seeking to expand operations particularly with regard to culinary herbs, which were introduced in 1999. The trade is entirely geared towards foreign markets, and currently Flex has a monopoly over the herbs export market in Uttarakhand.

Specifically, the herbs procured and subsequently exported by Flex include parsley, dill, basil, thyme, oregano, marjoram, chervil, sage, mint, savory, tarragon, rosemary, green pepper, lemon grass and coriander. These are in much demand in EU, US, Middle East and Canadian markets, and to gain entry and be able to sell to major wholesalers; Flex had to obtain necessary certification from EuropGap, ISO, BRC (British Retail Consortium). The stringent requirements imposed by these buyers imply that quality is a paramount consideration for cultivation. This has led Flex to adopt captive sourcing of raw materials, through both captive farming (for mushrooms) as well as near-contract farming through tacit agreements with farmers (for culinary herbs).

Flex works backwards, in that it conducts supply and acreage planning based on demands it receives for various quantities of culinary herbs from buyers. 80 per cent of its produce is sourced from Uttarakhand, and the remainder from adjoining districts in Uttar Pradesh. The sourcing area is divided into zones: Vikasnagar, Dehradun, Chiddrawala, Haridwar, Laxar, Najibabad and Saharanpur. Note that all these areas lie in the plains or the foothills. Based upon the orders received and the suitability of different zones for production of particular herbs (water availability is a crucial consideration), quantity targets are set for farmers, and rates fixed beforehand. They provide seeds and other inputs needed to farmers free in the first instance, and thereafter deduct the costs of these (the inputs are eventually provided on a nc profit-no loss basis) over 3 installments of payments made for the produce. Over the period of cultivation, the farmers are subject to constant monitoring and supervision. Field surveyors visit plot at least twice a week to maintain highest standards of quality. Transport for the ready crops is arranged and provided by Flex at the farm-gates.

Flex states that about 80 per cent of their farmers have land-holdings between 2 and 7 hectares, and 10 per cent each for those owning less than 2 and more than 7 hectares of land. Flex states that it prefers working with individual farmers rather than small groups, as groups may often try to manipulate and possibly hold the company to ransom. Groups and federations, they believe are more suited for the more remote areas with scattered plots, for the obvious advantages of volume and scale economies. Control over the quality and quantity of produce is critical, and hence heavily prioritised. In cases such as these where lucrative, high-value crops are grown under contractual terms (if not in explicit contracts *per se*) the possibility of side-selling is also very high. Flex has also come across the occasional case where farmers choose to sell to other buyers. This results in immediate snapping of all ties with the particular farmer, thus demonstrating the importance of trust and loyalty to others. Flex contends that even in the face of falling prices of particular crops, they stick to the agreed-upon prices and expected the farmers to show such integrity as well. In a genuine case of damage due to bad weather etc., Flex may write off the input credits given, but usually would not pay farmers any compensation as they believe the loss must be shared by both parties.

As a large player on the field, which has successfully demonstrated how the potential of niche product exports can be harnessed and built upon, Flex is an ideal case in point for illustrating the positive role good business models can play in the state's development process. However, they also mention certain shortcomings of the current policy environment and suggest changes. First they clearly state that poor road connectivity, irrigation and small land holdings may prevent the replication of initiatives such as theirs in the hilly areas of the state (despite the great potential they say may exists in the hills for particular herbs, especially oregano, sage and rosemary). Credit availability is a huge problem and discourages taking risk. Also, they mentioned that if a few credible local certification agencies were to come up, quality certification would not be as expensive as it currently is. In the years to come, Flex hopes to establish links with more and more Uttarakhand farmers to be able to procure, process and export higher volumes and varieties of herbs, fruits and vegetables.

V.8. Conclusions

Given Uttarakhand's natural biodiversity wealth and potential for herbal produce, policy should be directed at tapping into this herbal wealth in ways that best serve the economic interests of the mass of farmers and collectors alike. Policy must therefore take into account the advantages and pitfalls of promoting herbal cultivation on a large scale. There are a number of issues to be considered.

First, it is clear that there is a tension in the collection *versus* cultivation speace, that often both cannot compete effectively for the same crop and that the existing institutional and legal frameworks need to provide a conducive environment for herbal plants to come up in a significant manner in the state. For instance, excessive regulations over cultivation of certain crops only encourage corruption and connivance on part of the Forest Department to let illegal trade go unchecked. For species that are abundantly available in the wild and face no imminent threats of extinction, pushing an aggressive cultivation policy will not be an economically sound proposition. The Forest Department should be trained as to be neither unnecessarily high-handed while overseeing trade in sustainable species, nor being lax in addressing the important concerns of ecological balance and protection of endangered species.

There seems to be considerable confusion over obtaining permits for collection from the wild, as well as cultivation of specific herbs. While on the one hand local communities report that they are not allowed to collect herbal produce from the wild, there are ample suggestions that large scale illegal extraction and collection is simultaneously taking place. Giving local, and often traditionally forest-dependant communities the first right of preference when giving out such permits could be a step in the direction of equitable growth for the local economy.

Large volumes of the trade are done through illegal channels. This creates great uncertainty about prices, and may leave farmers at the mercy of local monopsonists who can easily take advantage of the lack of information. A right step that has been taken in this direction is the establishment of herbal *mandis* in the state. Although a number of problems may be associated with this channel of trade as well, it gives the farmers at least one systematic channel through which to sell their produce competitively, rather than through individual traders (In practice however, these decisions may be taken by the cooperatives and farmers' unions). Such creation of markets, coupled with more herbs-processing units that may be set up across the state offers the farmer/collector a number of options to choose from, and arguably a chance to optimise her sales' earnings.

Technical support regarding cultivation practices, processing, storage etc., is also rather weak at the implementation level. The case studies discussed above indicate that where there are tie-ups with buyers who are very keen and demanding of quality standards, they ensure that all good practices are disseminated and complied with. The state must encourage private investment in herbal products processing in the state, not only to boost industrial capacity, but also for the great potential for backward linkages, and associated benefits accruing to small farmers.

Development interventions in this sector could take on many roles. The establishment of more collection centres through the Forest Department, on perhaps a smaller scale than the existing herbal mandis but reaching the remote areas could be a welcome step towards offering the farmers a platform to aggregate, grade and sell their produce. This would also dissuade much of the illegal and informal trade that is rampant. Investments to provide accurate information to farmers and collectors on herbal prices are needed so that farmers are not exploited by traders. Catalyst programmes, which federate farmers and collectors and generate volumes (bargaining power) with buyback guarantees from processing companies, have been suggested as the means to move forward. These steps, along with greater investments in the extension services (currently collection and cultivation is carried out by relatively untrained labour), drawing from traditional knowledge systems as well as modern technology would be critical steps that the Uttarakhand herbal sector could immensely benefit from.

The question for the state is not whether or not to pursue agricultural diversification and income generation through herbal crops, rather it is, how it may do so, and under what institutional regimes. The particular benefits of large scale trade of certain crops may not be as high as others and there needs to be a case by case evaluation of what may be viable, or otherwise. It seems that many medicinal plants with long gestation periods and that grow in the wild cannot be viably cultivated. On the other hand the very visible success of culinary herb exports indicates that there may be very lucrative niche markets that some enterprise, creativity and sound planning can effectively tap into. The challenge lies in identifying those specific niches where strong market linkages may be forged, and in building the incentives, training and information-related support structures necessary for its sustainability.